The Making of Mankind

'WHAT MADE US WHAT WE ARE? WHAT WAS IT THAT, OVER MANY MILLIONS OF YEARS, TRANSFORMED PRIMITIVE HUMAN CREATURES INTO MODERN MANKIND? Humans have a deep desire to understand their origins. At no time in the past has it been so important for us to try to do just that.'

A decade of remarkable discoveries has made it possible for our origin as a species to be presented for the first time on the basis of scientific evidence. These discoveries and their wide-ranging implications form the core of THE MAKING OF MANKIND as a book and as a remarkable BBC TV series.

Although evolution fascinates many people there has until now been a wide gap between general curiosity and specialised knowledge about our distant past. Richard Leakey gives life to scientific data, stressing the excitement and challenge of 'the search for mankind'. He argues persuasively and clearly that the history of our mutual ancestors is not confined to the past: it throws light on the deepest questions of human nature and aspiration, and gives hope for the future.

Leakey's fascinating and vivid work spans seven continents, from the fossil beds of Kenya to the caves of Peking Man, and covers millions of years of evolution. It shows why the history of mankind is living history as compelling and urgent as history in the making.

'No one is better equipped for the task, other books cannot hope to compete with his practical account. Well written and beautifully illustrated.' Desmond Morris, author of *THE NAKED APE*.

Richard E. Leakey

The Making of Mankind

ABACUS

First published in Great Britain in 1981 by
Michael Joseph Limited
44 Bedford Square, London WC1

Abacus edition published by
Sphere Books Limited
30-32 Gray's Inn Road, London WC1X 8JL
1982

This book was designed and produced by
The Rainbird Publishing Group Limited
40 Park Street, London W1

House Editor: Linda Gamlin
Designer: Martin Bristow
Index: Michael Scherk
Production: Elizabeth Winder

Text photoset by SX Composing Limited,
Rayleigh, Essex, England

Colour origination by Gilchrist Brothers Limited,
Leeds, England

Printed and bound in Spain by Printer Industria Gráfica, SA
Sant Vicenç dels Horts-Barcelona D.L.B. 11136-1981

Contents

Foreword 6
1 Understanding Our Origins 9
2 Time and Change 22
3 Ape-like Ancestors 40
4 The Early Hominids 55
5 The Litter of the Past 76
6 Life as a Hunter-gatherer 97
7 New Horizons 110
8 The Birth of Language 127
9 Neandertal Man 142
10 Ice Age Art 160
11 Hunting in Transition 184
12 A New Way of Life 198
13 The Making of Human Aggression 219
14 The Future 238
Acknowledgments 248
Illustration Sources 250
Index 251
Glossary 255
List of Maps 256

Foreword

Since I first began my career in the study of human evolution, I have been continually impressed by the tremendous public interest that exists for the subject. I have enjoyed the chance to lecture to large audiences in places as far afield as Peking, Nairobi, Gaberone, Amsterdam, London, Paris, New York and Los Angeles. Wherever I have spoken, I have found the same response: a deep interest and a sincere curiosity about our origins. Many of the questions asked and the doubts expressed have been the same, regardless of the cultural or educational background of those I have spoken to.

I have tried to respond to these questions at a popular level. I hope that I can share the fascination of our evolutionary history with a great many, and perhaps provide a bridge between the serious professionalism of the scientists and the fundamental curiosity of us all. This book, and the television documentary series on which it is based, are an attempt to do that. In the book I have endeavoured to provide a written account of the issues and ideas that are covered in the seven-part series. Inevitably the written work is a more durable record and I hope that the two together will provide a lasting insight into what is surely an important dimension of humanity.

So much of our past and the events leading to our presence on this planet is now known and it makes a fascinating story. We can look at evolution and we can speak of facts, but above all we can rejoice in the knowledge that we are what we are because of a shared evolutionary past. To me, this is a tremendously exciting point to have reached, and I believe that much can be done to improve the future prospects of our species on the basis of this knowledge.

One of the problems of any narrative on a multidisciplinary subject is the difficulty of covering every aspect, of being up-to-date and of offering a totally objective account of the facts. I know that I may well have failed, and that some important discoveries have been glossed over rather too quickly. I know that there is an increasing body of evidence for the later stages of our prehistory, in areas such as Africa and Asia, that is not adequately represented between the covers of this book. I am painfully aware that the rate of progress in this field is so fast that new discoveries may change all our ideas about human evolution in the time it will take for the book to be printed. I regret this and can only hope that there will one day be an opportunity to tell even more of the story of our past.

As this book was nearing completion I learned of several important new pieces of evidence that should certainly have been part of the story. Perhaps the most interesting one is the discovery, at Saint-Césaire in western France, of the skeleton of a classic Neandertal that is dated at a substantially younger age than previous European finds of this type. This must have a major influence on any discussion about the position of the European Neandertals in the story of human evolution.

The whole field of human prehistory is fascinating and it is a science that is on the move. Each year, more discoveries are made and as a consequence our understanding changes. With each change we are closer to the truth and that, of course, is our objective.

In this book, I have written of many things and I refer to the work of numerous colleagues. I have attempted to mention as many as possible by name, but sometimes I may not have given sufficient credit where it is due. Often I am unaware of how ideas that are now familiar and widely accepted were originally developed. To all my colleagues I am grateful and I hope that this book and the television series will do justice to our collective endeavours.

1 Understanding Our Origins

The first serious decision I made about my life was that I very definitely was not going to follow in the Leakey tradition. My parents, Louis and Mary, were world-famous for their work and discoveries in the field of African prehistory. Their commitment to the quest was absolute, and it generated an atmosphere of unusual intensity to which I, like my brothers, Jonathan and Philip, was exposed from my earliest days. My supposedly resolute decision not to become involved in fossil-hunting was made not because I did not enjoy what it offered, but because I desperately wanted to be independent, to be free of the world that my parents were so closely associated with.

I succeeded for a while. By the age of sixteen I was doing very well as a supplier of small animals and animal skeletons to research institutions, and while still a teenager I established a photographic safari business. I also learned to fly, much to my parents' dismay. My 'freedom' seemed assured.

But slowly I was drawn back to the world of fossils. The fascination of what lay hidden in those ancient rocks gradually proved to be stronger than my rebellious wish to do something outside my parents' influence. I began occasionally to join in excavations at several different sites in Kenya. With no relevant academic qualifications behind me, my education consisted of what I saw and heard while I worked with others. Soon I realized that, like my father, I was infected with a powerful urge to know about the past, to know where the human animal came from, to understand our origins. I could no longer resist the tremendous challenge and excitement of the search for mankind.

'The ultimate concern'

Humans have many unusual characteristics, not least of which is our intense curiosity about our relationships with the world around us. We look at the many animals with which we share our planet and ask what it is that makes us different from them. We wonder if there are other living worlds scattered through the universe. Philosophical thought through the ages reflects an obsession with the question of what it means to be human.

One of the greatest television successes of the 1970s was *Roots*, the fictional story of a young American called Kunta Kinte who traced his history back through two centuries to a village in West Africa where his forebears had been taken into slavery. Social, economic and ethnic barriers vanished as huge audiences identified with the young man's need to reach back through time and find his origins. The story struck a chord with an apparently universal uncertainty about where we come from. The desire to remove that uncertainty is very strong indeed: it is what the great theologian Paul Tillich referred to as 'the ultimate concern'.

Throughout recorded history, religions and mythologies have developed

Richard Leakey at Koobi Fora in northern Kenya, examining stone tools made by our ancestors about half-a-million years ago.

to provide people with answers to their 'ultimate concern'. Each tribe, state and nation has explained its own creation, usually at the hand of an all-powerful god who subsequently peopled the rest of the world with rather inferior beings. The fact that all the world's cultures have some sort of god or gods is seen by many people as significant evidence for the existence of such a deity. Others, myself included, prefer to infer from it something about the special nature of the human mind.

The central importance of religion in most cultures is easily seen in the pages of history. But the human story stretches back well beyond formal records, which only began some 6,000 years ago. How long before this did the question of the 'ultimate concern' formulate itself in the human mind? Does the emergence of Ice Age art, some 35,000 years ago, show that spiritual quality? Can we say that philosophical questioning began with the evolution of anatomically modern people some 40,000 or more years ago? What, then, of those who went before? I believe it is likely that the urge to question one's origins has a very long history indeed, probably much longer than at first glance seems evident.

Although Darwin was careful not to be explicit about the origin of man, his theory of evolution clearly implied that humans beings were descended from an ape-like animal. Following the publication of *The Origin of Species* many newspaper and magazine cartoons reflected the general shock and alarm at the idea of being related to the apes. This cartoon appeared in *Punch* in 1861.

Western civilization has, for more than a millennium, been based on a Judeo-Christian religious foundation that views human beings, and the planet we inhabit, as the central focus of God's province. This comfortably egocentric view was challenged more than four centuries ago when Nicolaus Copernicus dared to suggest that the earth was not the centre of the universe, but only one of several planets revolving around a fairly small star. During the first three decades of the seventeenth century, first Kepler and then Galileo confirmed what Copernicus had suggested and the gathering storm broke, with both church and state condemning the astronomers for their heresy. Galileo was called before the Holy Inquisition and forced to renounce his findings, but the true nature of the physical universe became more and more apparent through the continued pursuit of science, and eventually the established view was overthrown. Our planet is indeed a humble body in a vast array of solar systems and galaxies. Science had begun its erosion of the Judeo-Christian view of the world.

The next major upset came in the middle of the nineteenth century when Charles Darwin published his book *The Origin of Species*. Darwin had restricted himself in the book to the following comment about the position of humans: 'Light will be thrown on the origin of man and his history.' In later editions he added the word 'Much' to the beginning of this sentence. But although Darwin was at pains not to be explicit about the origin of man, the clear implication of his theory of evolution was that humans were descended from some ape-like animal and were not the product of special creation. Nineteenth-century society was utterly shocked, and Darwin was roundly attacked by the church and by many scientists. It was intolerable that man should be supposed to be directly related to the common creatures of the world. The Judeo-Christian answer to the question of 'ultimate concern' was under threat, and those who relied on it were prepared to fight bitterly in its defence.

Following the publication of Darwin's theory, the wife of the Bishop of Worcester is said to have exclaimed: 'My dear, descended from the apes! Let us hope it is not true, but if it is, let us pray that it will not become generally known.' The lady's alarm was reflected in many newspaper and magazine cartoons at the time, some showing the apes equally surprised at having such unusual relations. Our very close biological relationship to apes has, of course, been confirmed scientifically in more recent times. For instance, biochemists have demonstrated that the proteins that make up the bodies of humans and chimpanzees differ in structure by less than one

Louis Leakey (left) with Raymond Dart in 1970.

per cent. Such biochemical similarity betrays a very close evolutionary relationship indeed, and I would go so far as to suggest that, were it not for our ego and concern to be different, the African apes would be included in our family, the hominidae.

What this means is not, as the Bishop of Worcester's wife feared, that we evolved from the chimpanzees and gorillas, but rather that, at some time in the past, we shared with them a common ancestor. At its simplest, one can imagine that the descendants of this common ancestor evolved along two paths, on the one hand producing the modern apes, and on the other giving rise to human ancestors, the hominids. But in fact, one should avoid thinking in terms of simple straightforward paths along which our ancestors progressed inexorably towards the present. From the prehistorian's point of view, it is useful instead to think of a stage on which many different players have played out a very long and complex drama. We want to know how many actors were involved in the drama, when they entered into the plot, how they were related to the other players, and when they finally departed from the stage.

The Leakey tradition

When, in 1926, my father began his search in Africa for evidence of human ancestry, virtually none of the players had been located, and the roles of those that had been found were widely misunderstood. At the time of his death in October 1972, Louis had spent almost half-a-century energetically taking part in a bid to understand more about our past. Together with Raymond Dart and Robert Broom, both of whom worked in South Africa, Louis made up a trio of really great men who pioneered the study of human origins in Africa.

Louis was in his early twenties when he decided to pursue a fossil-

hunting career. Until then, he had intended to follow his father's example and be a Christian missionary in Kenya. Although he was born and brought up in Kenya, he returned to England to be educated, first at a school in Weymouth, and then at St John's College, Cambridge. During his university years, Louis received an injury to his head while throwing himself with typical energy and enthusiasm into a game of rugby football. He suffered severe headaches as a result of this injury and his doctor recommended a rest. For his 'rest' he joined a British Museum expedition to Tanzania, in search of fossil dinosaurs. The experience was enough to convince him that fossil-hunting was what he wanted to do. 'Had it not been for my accident,' Louis wrote in his autobiography, 'I should certainly not have applied to go on this expedition, and I should therefore never have had the really practical training in methods of fossil collecting and preservation. . . . My luck had certainly changed in a most unexpected manner.' So, when he returned from the British Museum expedition, he finished his degree in anthropology and began preparations for returning to Africa with a venture of his own in 1926.

Louis's decision to look for early man in Africa was considered to be utterly misguided by the academic establishment. Although Darwin had suggested in his 1871 book *The Descent of Man* that human forebears might be found in Africa, European scholars in the 1920s considered their own part of the world to be a much more likely location. It was thought that a perfectly respectable ancestor, 'Piltdown Man', had been discovered in England itself: only later was the Piltdown skull shown to be a forgery. The so-called 'man-ape', a genuine hominid discovered by Raymond Dart in South Africa in 1924, was widely dismissed as being more ape than man.

Nevertheless, Louis went to Africa. He excavated at many sites in Kenya and Tanzania, some very ancient and some relatively recent. His favourite site, however, was Olduvai Gorge. Hans Reck, a German geologist, had worked at the Gorge in the years before the First World War, and he had recognized the importance of this fossil site. In 1931 Louis discovered well-worked handaxes, probably about a million years old, lying on a slope: his confidence in Olduvai had been vindicated, but almost thirty years were to pass before an ancient hominid fossil was to be found there.

During those three decades Louis was extremely busy in and around the National Museum in Nairobi, as well as working at many other sites in East Africa. Because of these commitments and a lack of funds Louis and my mother, Mary, were only able to visit Olduvai infrequently, hence the frustratingly slow progress. Jonathan, Philip and I often accompanied our parents on their visits, both to Olduvai and to other sites. Not only did we absorb a lot of information about the techniques of searching for fossils and excavating them, but we also learned about survival in rough country: how to obtain food, where to look for water in dry terrain, and which plants are useful medicines. It was an education that helped instil in me a great feeling for East Africa and the animals that live there.

Ironically, although Jonathan, my eldest brother was the only one of us to find a fossil of any significance during our younger years, he was also the only brother to withdraw completely from fossil-collecting. Until recently, Philip, the youngest, occasionally helped Mary with her work at Olduvai Gorge and the nearby site of Laetoli. However, his convincing election to the Kenyan parliament will probably restrict his excursions into prehistory at least for the time being.

By the mid-1960s Louis and Mary had amassed an impressive list of fossil finds from which some details of human prehistory could at last be inferred. Louis believed, for instance, that some two million years or so ago several

Olduvai Gorge in Tanzania. The Maasai gave it the name *Ol duvai* meaning 'the place of the wild sisal', after the spiky plants which can be seen growing on the right of the picture. Louis Leakey began his work at the Gorge in 1931, and excavations are still being carried on there today by Mary Leakey and her team.

different species of hominids existed, some of which eventually became extinct while one of them, which he named *Homo habilis*, ultimately gave rise to modern humans. He also believed that the origin of the *Homo* line went even further back in time. Such an extensive ancestry for the human stock was a very new idea at the time, and one which many researchers could not bring themselves to accept.

My first major involvement in fossil-hunting was in 1967 when I took charge of the Kenyan contingent of an international expedition to the Lower Omo Valley in Ethiopia. The expedition also included teams from France and the United States. Although the Omo Valley area was promising, I decided on another location for my own future research. This came about quite by chance when I had to return briefly to Nairobi during the work at the Omo. On the return journey, in order to avoid a storm, we flew over the east side of Lake Turkana (formerly Lake Rudolf) rather than take the more normal westerly route. I was not piloting the plane so I had an opportunity to look closely at the terrain below. It seemed to me that what many people had taken to be dark-coloured lava stretching mile after mile towards the distant hills might in fact be sandstone and other sedimentary deposits. If this were the case it *might* prove to be rich in fossils. A quick ground-check by helicopter some days later established that there were indeed fossils to be found.

Soon afterwards, I approached the National Geographic Society for a grant of 25,000 US dollars to support a preliminary expedition to this new area. My father made it quite clear at the time that he was against the new venture and this response made me even more determined. I have often wondered why he reacted in this way. I think that perhaps he was really keen on the idea, and, knowing how I generally reacted to him, *appeared* to object to the plan with the very purpose of making me determined to see it

Richard Leakey with Kamoya Kimeu during the 1969 expedition to Lake Turkana. To help exploration in the rugged country, camels were used for transport, but they proved unco-operative and difficult to control.

through. The National Geographic Society took a chance and gave me the grant, thus initiating an astonishingly rewarding period of my life.

I was lucky to have Kamoya Kimeu with me on the preliminary exploration of the lakeshore in 1968. Kamoya had worked with Louis and Mary at Olduvai since 1960, and he had become an expert prospector for fossils. We had been on other expeditions together, and he was pleased to join the Lake Turkana party. Also on the expedition were John Harris, Bernard Wood, Paul Abell, Bob Campbell and my former wife, Margaret. Fossils simply littered the landscape. We found three hominid jaw fragments, and these firmly established the potential for further work at this new site.

The eastern shore of the lake ultimately proved to be formed of layer upon layer of sandstone which, to my great good fortune, contain a rich store of fossils. In the years since the first excursion to Lake Turkana, the large team of researchers working at the site has unearthed remains of more than 160 hominids. Together with a number of very important discoveries in Tanzania, Ethiopia and South Africa, the finds at Lake Turkana are beginning to provide a fairly detailed picture of our past. Admittedly, the picture of human origins is still sketchy in some respects, but important outlines can be discerned and, if progress is maintained at the same pace for the next decade, I believe that a fairly comprehensive picture will finally emerge.

I was able to return to Lake Turkana for three months in the summer of 1969, and during this expedition we used camels as pack animals and for transport. The other members of the team were Kamoya, Peter Nzube and Meave Epps, who was later to become my second wife. On one occasion we had planned to extend our survey right up to the border with Ethiopia, but a remarkable discovery sent us scurrying back to base camp. We had left our camels and temporary camp in the cool of the dawn and set out to walk among the fossil-beds, looking for interesting specimens. By 10.00 a.m. we had reached the furthest extent of the fossil deposits in that particular direction. Meave and I were thirsty, so we headed back in a more or less direct line to where we thought the camels were. While I was walking down a dry riverbed my eyes fell on something that made me stop in my tracks: it appeared to be a hominid cranium sitting on the sand. We advanced to the spot to find the ancient bony face of an intact hominid skull staring at us. It was a truly extraordinary moment.

The cranium had clearly been embedded in the bank of the stream. Rain and the flow of water along the stream had gradually been eroding the bank away, and the cranium had probably rolled out during the last heavy rainstorm. If we had not found it then, the seasonal river would have soon washed it into oblivion. By coincidence the cranium was of the hominid type that my mother had discovered at Olduvai almost exactly ten years earlier. By a further coincidence, Mary happened to be visiting the base camp at Koobi Fora and so we rushed back, hoping that she would not have departed for Nairobi before we got there. It was a two-day camel ride to the camp but, luckily, she was still there when we arrived and was thrilled to see the magnificent specimen. I should say that this kind of discovery is very much the exception. Most finds are the result of many patient months of work, searching in likely deposits of rock.

Our first camp at Lake Turkana was a temporary affair: a small collection of tents and an open-sided thatched structure in which to work during the heat of the day. In 1969 we began building more permanent *bandas*: thatched huts with stone floors. The camp is now relatively large, capable of housing some sixty people at times. Research teams come from several countries each year to participate in many different projects at the lake. If the work at the lake during the past decade has been successful, it is

The 1969 expedition resulted in a remarkable discovery: an intact cranium of the hominid species known as *Australopithecus boisei*. It was almost exactly ten years since Mary Leakey had discovered a cranium of this same species at Olduvai Gorge.

because of the extensive co-operation and team spirit of my many colleagues, to whom I will always be grateful.

With the excitement of the 1969 season still running high, Meave and I returned to the site periodically to continue the prospecting and, in fact, simply to be at Koobi Fora. We would leave Nairobi at midday on Friday, drive through the night to Ferguson's Gulf on the west side of the lake, and then cross to the east side in our small boat on Saturday morning. Saturday and most of Sunday were spent looking for fossils or walking, and then we would cross back to Ferguson's Gulf before sunset, driving again through the night, to be back in Nairobi by Monday morning. We enjoyed it enormously, for Lake Turkana is a truly magical place, the stark fossil-beds broken here and there by swathes of green and punctuated by the delicate pink of the desert rose. But although these weekend trips were enormous fun, we realized after four or five months that they were too exhausting. We soon established a permanent lakeside camp, manned by staff throughout the year, so that we, and others, could visit Koobi Fora more easily.

In a very real sense, each hominid specimen that is found at Lake Turkana is just as important as any other: each one forms part of a gradually emerging story. Inevitably, though, some receive more attention than others, perhaps because of certain special characteristics, or because, when they were found, they changed earlier ideas about our evolutionary history. One such example is a cranium known as 1470, which Bernard Ngeneo, one of Kamoya's team, found mid-way through 1972. Bernard happened to look more attentively at an area which many people must have passed repeatedly. He saw some small bone fragments which he recognized as part of an unusually large hominid cranium. It was a hominid that lived a little more than two million years ago. Unlike the first cranium found near Koobi Fora, this one was in hundreds of pieces, and many weeks of patient jigsaw work by Meave and later by the anatomist Alan Walker were required to piece it together.

As the reconstruction progressed, excitement steadily rose in the camp because it became clear that the early suspicions about the cranium were correct: this ancient hominid had a remarkably large brain, and yet it had lived such a very long time ago. This cranium was of the same type as the one that Louis had named *Homo habilis*, although 1470 was a little older and more complete than the specimens from Olduvai.

When the restoration of the cranium was almost finished, I took it back

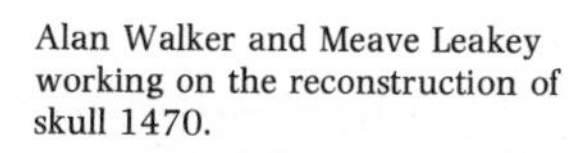

Alan Walker and Meave Leakey working on the reconstruction of skull 1470.

to the museum in Nairobi, and it was there that Louis examined it shortly before leaving for his final trip to Europe and America at the end of September. Naturally he was delighted. Skull 1470 seemed to confirm what he had proposed years before, that the *Homo* line was much older than had been supposed. He knew there would be sceptics, and he said almost mischievously, 'They'll never believe you!' Shortly after arriving in England, Louis died of a heart attack.

Homo habilis may well have played a key role in the unfolding drama of human origins. But, as I have already said, the search for our origins consists of far more than simply identifying the characters in the play: we need to know what they did, when they arrived on the stage, and when and why they departed. In more specific evolutionary terms, the question that Louis was asking and the question that I and my colleagues continue to ask is: what was it that, over many millions of years, transformed primitive pre-human creatures into modern mankind? What made us what we are? The whole answer, it has to be admitted, still eludes us, but the many important discoveries and the development of new ideas during the past decade have brought us much closer to being able to say what made us 'human'. It is the exciting finds of the 1970s, together with the opportunity that they give us to understand ourselves, that I want to describe in this book.

The human animal

What are we? To the biologist we are members of a sub-species called *Homo sapiens sapiens*, which represents a division of the species known as *Homo sapiens*. Every species is unique and distinct: that is part of the definition of a species. But what is particularly interesting about our species? For a start, we walk upright on our hindlegs at all times, which is an extremely unusual way of getting around for a mammal. There are also several unusual features about our head, not least of which is the very large brain it contains. A second unusual feature is our strangely flattened face with its prominent, down-turned nose, Apes and monkeys have faces that protrude forwards as a muzzle and have 'squashed' noses on top of this muzzle. There are many mysteries about human evolution, and the reason for our unusually shaped nose is one of them. Another mystery is our nakedness, or rather *apparent* nakedness. Unlike the apes, we are not covered by a coat of thick hair. Human body hair is very plentiful, but it is extremely fine and short so that, for all practical purposes, we are naked. Very probably this has something to do with the second interesting feature of our body: the skin is richly covered with millions of microscopic sweat glands. The human ability to sweat is unmatched in the primate world.

So much for our appearance: what about our behaviour? Our forelimbs, being freed from helping us to get about, possess a very high degree of manipulative skill. Part of this skill lies in the anatomical structure of the hands, but the crucial element is, of course, the power of the brain. No matter how suitable the limbs are for detailed manipulation, they are useless in the absence of finely tuned instructions delivered through nerve fibres. The most obvious product of our hands and brains is technology. No other animal manipulates the world in the extensive and arbitrary way that humans do. The termites are capable of constructing intricately structured mounds which create their own 'air-conditioned' environment inside. But the termites cannot choose to build a cathedral instead. Humans are unique because they have the capacity to *choose* what they do.

Communication is a vital thread of all animal life. Social insects such as termites possess a system of communication that is clearly essential for their complex labours: their language is not verbal but is based upon an exchange

Opposite: Once restored, it became obvious that skull 1470 was of the same type as one from Olduvai Gorge which Louis Leakey had called *Homo habilis*. The new find delighted Louis, for it seemed to confirm his theory that the *Homo* line was much older than had been thought.

of chemicals between individuals and on certain sorts of signalling with the body. In many animal groups, such as birds and mammals, communicating by sound is important, and the posture and movement of the body can also transmit messages. The tilting of the head, the staring or averted eyes, the arched back, the bristled hair or feathers: all are part of an extensive repertoire of animal signals. In animals that live in groups, the need to be able to communicate effectively is paramount.

For humans, body language is still very important but the voice has taken over as the main channel of information-flow. Unlike any other animal, we have a spoken language which is characterized by a huge vocabulary and a complex grammatical structure. Speech is an unparalleled medium for exchanging complex information, and it is also an essential part of social interactions in that most social of all creatures, *Homo sapiens sapiens*.

All the points I have mentioned are characteristics of a very intelligent creature, but humans are more than just intelligent. Our sense of justice, our need for aesthetic pleasure, our imaginative flights and our penetrating self-awareness, all combine to create an indefinable spirit which I believe is the 'soul'. Like all animals, we have to concern ourselves with the business of survival, of obtaining food and shelter, but that is not all. As Dostoevsky wrote: 'Man needs the unfathomable and the infinite just as much as he does the small planet which he inhabits.'

Our place in the universe

The physical universe has existed for some 20,000 million years but our tiny corner of the universe, the solar system, is a relative youngster within that great expanse. The earth was formed just 4,600 million years ago. Life on earth arose surprisingly soon after the hot gases and rocks condensed to form our planet, probably about 3,500 million years ago. The first living organisms were small, simple cells, very like bacteria, and they remained unchanged for several thousand million years. Then larger and more complex animals, like worms and jellyfish, began to appear. Life flourished in the seas at first and it was not until 400 million years ago that dry land was successfully colonized. Since that time several hundred million animal species have arisen, most of which later became extinct. Of all the animal species that ever existed, only one per cent are alive now: clearly, the ultimate fate of most species is extinction. Some survive for perhaps a few million years, some for much longer, but the chances are that most species which come into being will ultimately vanish from the face of the earth.

What about our species? So far we have been around for a mere 100,000 years or so. Our immediate forebears, *Homo erectus*, appear to have lasted for about one-and-a-half million years, and before that *Homo habilis* occupied parts of Africa for almost a million years. In theory, it seems that our prospects are good for the next million years at least. Indeed, the latest representative of the *Homo* line, *Homo sapiens sapiens*, is able to exert so much more control over the environment than any other species that we should expect to be able to avoid the fate of extinction altogether. Humans are certainly extremely adaptable creatures and can respond to changes with appropriate technological solutions. Our chances of surviving would seem very good indeed. However, a glance at the pages of human history, documenting the events of the past 10,000 years, should serve to dull that note of optimism.

The fruits of the so-called Agricultural Revolution of 10,000 years ago nurtured the growth of world population and of concentrated population centres. As towns grew into cities and cities into states, human beings gradually turned to conflict and confrontation. Today we see a world in which human adaptability and technological wizardry has allowed occupation of virtually every part of the globe, and yet the future security of that world is under very real threat from the nuclear weapons of the superpowers. It is a sad irony that the same ingenuity and inventiveness that has breathed life into the most sterile parts of the earth could also turn the planet into one vast desert, where virtually nothing would survive because of nuclear fallout.

There is a popular notion that such a state of affairs is an inevitable consequence of human nature. *Homo sapiens sapiens*, the argument runs, is an innately aggressive creature, a 'killer-ape', and as technology becomes ever more refined, the military machinery will grow in sophistication and destructive power, and it *will* be used. If this argument is correct, then there appears to be very little one can do about it, for the holocaust will come sooner or later. But I am convinced that it is not correct, and that this popular notion of the 'killer-ape' is one of the most dangerous and destructive ideas that mankind has ever had.

I began this chapter by suggesting that humans have a deep desire to understand their origins. It seems to me that at no time in the past has it been so important for us to try to do just that. The question of the 'ultimate concern' now has bearing on our continued survival, not through the next million years or even a mere thousand years, but simply to the end of this century.

Following a decade of remarkable discoveries in palaeoanthropology, our origin as a species can, for the first time, be presented on the basis of scientific evidence. Thus there is some hope that all people, regardless of creed or culture, will learn of the unity of mankind and of our shared past. More than a decade ago, the great Russian geneticist, Theodosius Dobzhansky, made the following comment: 'The relevance of biology and anthropology is evident enough. In his pride, man hopes to become a demigod. But he still is, and probably will remain, in goodly part a biological species. His past, all his antecedents, are biological. To understand himself he must know whence he came and what guided him on his way. To plan his future, both as an individual and much more so as a species, he must know his potentialities and his limitations.' My aim in this book is to bring together the results of important work, carried out by many scientists in many parts of the world, that throw light on man's biological nature. My hope is that this will help us to avoid a desperately frightening future.

2 Time and Change

If one looks back through any sequence of ancient fossil-bearing rocks, one notices a persistent theme: change. New species appear, while established ones vanish, but instead of showing a sequence of smooth transitions, the fossil record gives an impression that the changes happened in a series of jumps. It is as if one were looking through an album of photographs, rather than viewing a cine film of the past.

The mechanisms by which species arise and subsequently slip into extinction have puzzled biologists for a very long time. The problem first arose when amateur geologists and naturalists discovered the layered nature of certain rocks and the existence, within these rocks, of remains of animals that no longer exist. These discoveries implied that the living world had changed in a way that was difficult to reconcile with the concept of the Creation.

The so-called Diluvial Theory came to the rescue. This proposed that the extinct species had been victims of Noah's Flood. However, it became apparent that rocks often contained not one, but many layers of different extinct creatures, and this meant that a single flood could not have been responsible for their demise. The famous French geologist and naturalist Baron Georges Cuvier offered a solution with his Catastrophe Theory. The world, he suggested, had passed through a series of creations, each of which was followed by a global destructive event that wiped out most, if not all, of the earth's inhabitants. Twenty-seven such events had been calculated at the time of Cuvier's death in 1832.

The Catastrophe Theory appeared to account for the layers of fossils in the earth's rocks, but the length of time during which these events were supposed to have occurred was much too short. In 1650 James Ussher, the Archbishop of Armagh, had published his calculation of the age of the earth, based on information in the Old Testament. The date Ussher calculated for the Creation was 4004 BC, which gave our planet a very short history indeed. Cuvier and his followers reassessed the age of the earth and they decided that 70,000 years was about right. But the revelations of geology continued to attack the foundations of the conventional Western view of the world. James Hutton, a Scot, had created the basis of a 'new geology' during the eighteenth century, and Charles Lyell continued the revolution in the nineteenth century. According to the evidence he saw in the rocks, the earth was subject to very slow but steady formative processes. Lyell argued that not only did these processes take a very long time, but that they were also still acting. In other words, the world that people thought to be stable was instead undergoing continuous dynamic change.

Lyell published the first volume of his *Principles of Geology* in 1830, and this firmly established the new view. The world was soon accepted as being of extreme antiquity, not thousands but many millions of years old, and this

The Omo Valley, Ethiopia. The layered appearance of these hillsides is characteristic of sedimentary rocks, which are formed as silt is deposited by a lake or river. Animal bones that are covered by the deposits may become fossilized, and occasionally softer materials, such as leaves, are also preserved. If a river later cuts through the sediments, as the Omo River has done here, the layers of fossil-bearing rocks become exposed.

provided the essential geological background against which a theory of the slow evolution of species could be formulated.

The theory of evolution

Many people had puzzled over the origin of species, and some had come very close to the theory with which Charles Darwin is associated. One of these was his grandfather, Erasmus Darwin, a physician, poet and philosopher. In his later writings he mused over the possibility of all creatures sharing a common ancestor, and he speculated on the way in which species might be transformed. A more famous proponent of evolution was Jean-Baptiste de Lamarck, who published his theory in the year of Darwin's birth, 1809. Lamarck's theory enjoyed a brief spell of popularity, but it soon became discredited, largely because the influential Cuvier criticized it so strongly.

Lamarck mistakenly believed that evolution occurred because characteristics acquired in an animal's lifetime were passed on to its descendants. This would mean that giraffes had long necks because their forebears continually stretched their necks in order to reach the upper branches of tall

Charles Darwin as a young man.

trees. This is as absurd as supposing that if a man has had a leg amputated all his sons and daughters will only have one leg. For the most part, however, Lamarck's ideas were sound, and criticism of him has been unfairly hostile. It is worth noting that Darwin, too, believed that acquired characteristics could be inherited, but he did not *base* his theory on this mechanism as Lamarck had done.

Charles Darwin was the son of a formidable country doctor, Robert Waring Darwin, who was also a deeply devout man. It was intended that Charles would continue the family tradition, but while at medical school he almost fainted at the sight of surgery and so the church was thought to be a respectable substitute. Charles Darwin went to Cambridge to read divinity, but once again he failed to live up to expectations, preferring a life of hunting and socializing to serious study. He did, however, develop a passion for nature, with the encouragement of a Cambridge professor, J S Henslow. It was this last turn of events that eventually prevented Charles Darwin from becoming, as his father predicted, 'an idle sporting man'.

Darwin was not an outstanding student, but he was, in his uncle Josiah Wedgwood's words, 'a man of enlarged curiosity'. It was this quality, a recommendation from Henslow and the fact that he was a 'gentleman' that secured him a position on board H.M.S. *Beagle*, under the captaincy of Robert Fitzroy. Fitzroy's task was to map many poorly known waters that were important to the British Navy, an undertaking that was to involve a five-year circumnavigation of the world, starting in December 1831. Darwin's duty was to provide the companionship of a gentleman for the captain, although he officially went on board as the 'ship's naturalist'.

Despite repeated bouts of both seasickness and homesickness, Darwin was a highly enthusiastic 'ship's naturalist' and he amassed enormous collections of rocks, fossils, animals and shells before returning to England in October 1836. He kept a voluminous notebook of his observations, and he thought deeply about everything he saw. Darwin was at least as interested in geology as he was in the living world, and the combination of the two sciences proved highly productive. It was the sights he saw in South America and particularly on the Galapagos Islands that convinced Darwin that species could indeed change, that they were not immutable. In an account of the entire journey, Darwin wrote this of the Galapagos Islands: 'Hence, both in space and time, we seem to be brought somewhat near to that great fact – that mystery of mysteries – the first appearance of new beings on this earth.'

Darwin began his notebook on 'The Transmutation of Species' shortly after returning from his great voyage, as he recalled in a letter written in 1877: 'When I was on board the *Beagle* I believed in the permanence of species, but as far as I remember, vague doubts occasionally flitted across my mind. On my return home in the autumn of 1836, I immediately began to prepare my Journal for publication, and then saw how many facts indicated the common descent of species, so that in July 1837 I opened a notebook to record any facts which might bear on the question. But I did not become convinced that species were mutable until, I think, two or three years had elapsed.' In 1838, Darwin read an essay on populations by Thomas Malthus, and this sowed seeds that were to be important for the later development of his theory of natural selection. By 1842 he was ready to sketch an outline of his emerging ideas, and he extended this in 1844.

But more than twenty years elapsed between Darwin's return and the publication of *The Origin of Species* on 24 November 1859. For some reason yet to be fully explained Darwin 'sat on' the most revolutionary biological theory of all time, apparently reluctant to make it public. During those twenty years he collected more and more evidence by reading widely, by patiently watching life in the country around him, and by carrying out experiments. Eventually he was virtually forced to publish. Another English naturalist, Alfred Russel Wallace, sent him a brief paper for his comments and when Darwin read it, he found to his horror that it neatly summed up his own theory. As a response to Wallace's approach, Darwin was urged by his friends to deliver a joint presentation on the subject to a meeting of the Linnean Society in London in 1858. *The Origin of Species* appeared the following year. Darwin's apparent intention was to produce a monumental work of which *The Origin of Species* was merely 'an abstract', but in fact he never did so.

Selection and survival

The import of *The Origin of Species* is fourfold. Firstly, Darwin saw the living world as changing rather than static: species gradually become changed, so that new species emerge while other species fall into extinction. Secondly, Darwin believed this process to be gradual and continuous, and not to involve 'jumps' or sudden changes. Thirdly, Darwin postulated the idea of common descent: all mammals, for instance, share a common ancestor, as do all reptiles, all birds, all insects, and so on. Darwin even speculated that all life, including both plants and animals, might ultimately have come from a common ancestor. Fourthly, the mechanism of change, of evolution, was through natural selection, by which process those individuals with improved characteristics leave behind them the most offspring.

Natural selection can only operate if there is 'variation'. 'Variation' means that although offspring might display many characters in common with their parents they are never identical to them. There might, for instance, be variations in size, colour, muscular co-ordination and other structural or behavioural characteristics. Offspring are never exactly the same as each other, except in the rare cases of identical twins. There is a noticeable degree of variation both within families and within the population. We have an advantage over Darwin in that we can understand how variation occurs at a molecular level. We know that characteristics are passed from parent to offspring by chemical units which we call 'genes'. Variation is largely due to a reshuffling of the genes when the genetic instructions from the two parents come together.

Natural selection acts on variation by 'favouring' some animals over others. Animals survive in the world by being able to obtain sufficient food for their needs, as well as water, adequate shelter and so on. They must also be able to escape predators and avoid other hazards. This is a rule that applies throughout the living world. Furthermore, to produce offspring, they must successfully engage in mating, giving birth, and caring for the young. It is in this aspect of life that the selection process operates, for all individuals produce more offspring than will eventually survive. Some youngsters are just plain unlucky and fall victim to a predator at an early age, but overall, the ones that survive do so because they are best fitted to their surroundings. Because of variation, it follows that some individuals are likely to be more capable than others. It is these, the 'fittest', who survive, who thrive, and who leave most offspring of their own behind them.

Such a view of life seems, at first sight, to be cold and calculating. The term 'survival of the fittest', which Darwin used, was in fact coined by Herbert Spencer in his description of economically stratified society in the early nineteenth century. 'If they are sufficiently complete to live, they *do* live, and it is well they should live,' wrote Spencer. 'If they are not sufficiently complete to live, they die, and it is best they should die.' It is therefore not surprising that the laws of biology which Darwin formulated were invoked to support the social and political attitudes associated with *laissez-faire* capitalism. Indeed, the movement came to be known as Social Darwinism, something of which Darwin himself was very suspicious. Pronouncements such as 'The strongest and best survive, that's the law of nature after all, it always has been and always will be' or 'millionaires are the product of natural selection' become commonplace.

Strangely enough, Darwin's theories also attracted those of quite opposite political beliefs. Karl Marx was a great admirer of Darwin and was moved to comment on how much the author of *The Origin of Species* had drawn on the picture of life as he saw it in his own society. In a letter to his colleague, Friedrich Engels, in 1862 he wrote, 'It is remarkable how Darwin recognizes among beasts and plants his English society, with its division of labour, competition, opening up of new markets, "invention" and Malthusian "struggle for existence".'

Frequently overlooked, however, is the following passage from *The Origin of Species* which is clearly meant to set a different tone: 'I . . . use the term Struggle for Existence in a large and metaphorical sense, including dependence of one being on another, and including (which is more important) not only the life of the individual, but success in leaving progeny. Two canine animals in a time of dearth may be truly said to struggle with each other which shall get food and live. But a plant on the edge of a desert is said to struggle for life against the drought, though more properly it should be said to be dependent on the moisture.' Darwin was distressed by the political interpretations of his theory, when all he was concerned with was describing the natural world as he saw it. It was, perhaps, somewhat unfortunate that he used terms which could be seized on so readily and applied to human society by people who did not truly understand their meaning.

In order to survive, animals must be able to escape predators as well as find enough food and water for themselves. Some animals escape predators by stealth, by camouflage or by retaliatory behaviour, but others, such as these impalas, rely on rapid flight. Among such species the faster animals will tend to survive and breed while the slower ones will more often be killed. Thus natural selection 'favours' the faster animals.

Evolutionary success is not just a question of surviving: an individual must also pass on as many of its genes as possible to the next generation. Mating, giving birth and rearing the young are, therefore, vital activities.

Natural selection is generally a conservative force, ensuring the maintenance of certain physical and behavioural standards within a species. Under normal circumstances, the weak and the aberrant individuals pass on few, or none, of their genes. In a stable environment, therefore, species remain relatively unchanged for very long periods of time, but, as Darwin had observed in breeding experiments, animals can change, and this also occurs in the natural world.

Variation, as already described, arises largely through the reassortment of the parents' genes during reproduction, but there is also another source of variation: spontaneous changes in the genetic material sometimes occur. Genetic changes of this sort, known as mutations, are rare. The genetic material is surprisingly resistant to change and there are many complex chemical mechanisms that can correct mistakes. Nevertheless, mutations do arise occasionally. Their effect on the organism is random: most often they are harmful and the mutant individual does not survive, sometimes they are neutral and make no difference to their bearers, but very occasionally they are beneficial, in which case the origin of a new species is possible. What is not random is the environment into which mutants are introduced. The availability of fruit, leaves and potential prey, whether or not water is plentiful: factors such as these make up the forces of selection that will favour certain new physical or behavioural characteristics that may appear.

If one looks at the history of life on this earth, one detects a steady increase in the complexity of living organisms. It is often said that there is a progression from 'lower' to 'higher' life forms. Darwin disliked the use of these terms, as they imply inferior and superior positions in some kind of hierarchy. Instead, it is much better to view species, however simple or complex they may be, as being fitted to a particular kind of lifestyle in a particular environment. On the face of it, a limpet may seem to be leading a less interesting and less demanding life attached to its tideline rock face than, say, a lion. But in biological terms, both are very well adapted to the way in which they make a living. One is not better than the other, just different.

The growth in anatomical and behavioural complexity through time does invite the suggestion that some kind of guiding hand is at work in the process of evolution, moving the course onward and upward to the ultimate pinnacle, the human species. The apparent progress in evolutionary products is, however, just what one would expect from the action of natural selection. Ernst Mayr, one of the great evolutionary biologists of our time, explains it in this way: 'Evolution . . . is recklessly opportunistic: it favours any variation that provides a competitive advantage over other members of an organism's own population or over individuals of different species. For billions of years this process has automatically fuelled what we call evolutionary progress. No programme controlled or directed this progression: it was the result of the spur-of-the-moment decisions of natural selection.'

The interaction between the random changes in genetic material and long-term environmental changes has generated a living world of amazing diversity. It is a diversity of which humans are a part, not the pinnacle. In one sense, we are the product of a series of chance events; in another, we are the result of a progressive, but not purposeful, series of innovations.

One factor that helped to form Darwin's ideas on the origin of species was observation of the changes in animals and plants produced by artificial selection and breeding. Indeed, Darwin had spent a good deal of time breeding pigeons, and references to pigeons appear repeatedly in his book. A second set of observations that contributed to his theory concerned the

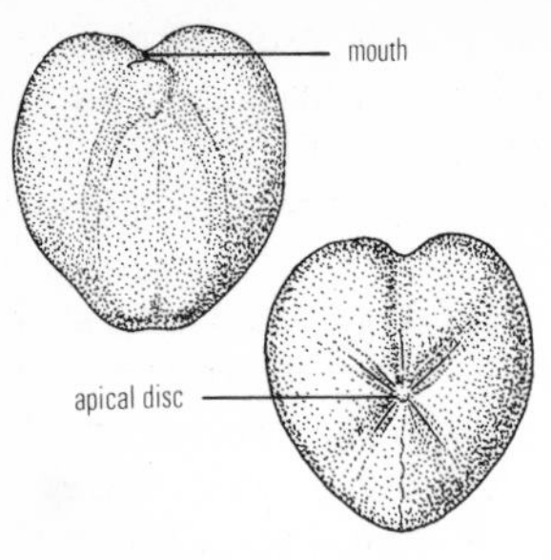

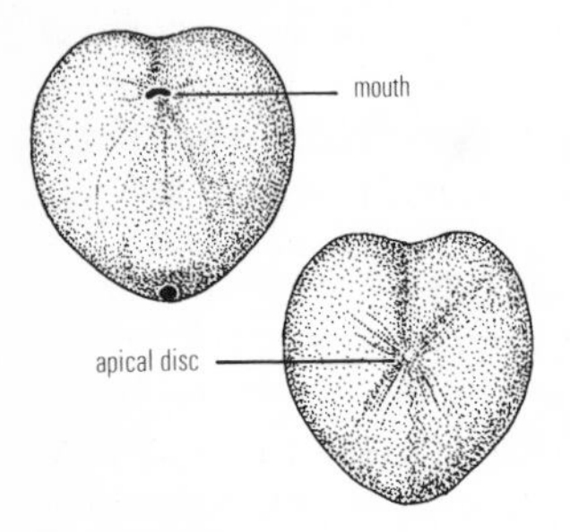

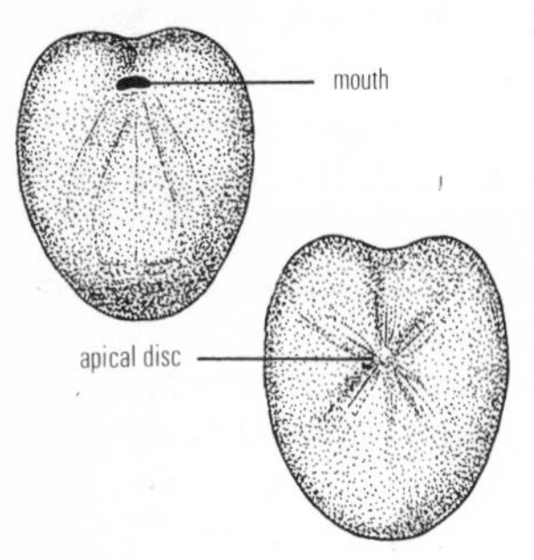

These heart-urchin fossils form a series in which the shape gradually changes, as does the position of the mouth on the lower side (left) and the apical disc on the upper side (right). However, there are very few such examples of smooth, gradual transitions to be found in the fossil record. This was a problem which greatly concerned Darwin, and it is at the centre of one of the most interesting debates in evolutionary biology today.

geological evidence for extinct species. He was impressed by the fossils entrapped in ancient sedimentary rocks, and, speaking of the obvious similarity between the living and the extinct species in South America, Darwin said, 'This wonderful relationship between the dead and the living will, I do not doubt, hereafter throw more light on the appearance of organic beings on our earth and their disappearance from it than any other class of facts.'

Darwin saw in the fossil record glimpses of past eras which, he inferred, were linked by steady and continuous change with each other and with the present. He envisioned a steady unfolding of new characteristics, as species gradually transformed from one into another, over a very long period of time. Though Darwin adduced the fossil record as support for his theory, the evidence to be seen there is more of an embarrassment than a prop. The record shows very few smooth transitions from one species to the next. Instead, new species seem to arise rather abruptly. There is a 'jerkiness' in the record that discomforted Darwin. He acknowledged this in *The Origin of Species*: 'Why then is not every geological formation and every stratum full of such intermediate links? Geology assuredly does not reveal any finely graduated organic chain; and this, perhaps, is the gravest objection which can be urged against my theory.'

The evolution of new species

This brings us to one of the most interesting contemporary debates in evolutionary biology. Darwin explained the 'jerkiness' of the fossil record by saying that the record was incomplete. If only one could collect fossils that represented more fully the passage of time, he implied, then one would see the transitional forms between species. An alternative explanation, put forward in recent times by American biologists Niles Eldredge and Stephen Jay Gould is that new species arise not as a consequence of slow, smooth transitions but as the result of relatively sudden events. In evolutionary terms, however, 'sudden' means hundreds or thousands of years.

Darwin's notion was that new species come about through the gradual addition of new features to an existing species, so if one examined the population at one point in time one would see the full characteristics of the parental species, while subsequent examination, perhaps a million years later, would reveal a related but distinct species, displaying new features. And at any point in between, there would be transitional stages with the new characteristics still incompletely developed. The evolutionary transition, he argued, ran throughout the whole species population. This theory has been given the imposing name of 'phyletic gradualism'.

'Gradualism has always been in trouble,' says Stephen Jay Gould, 'particularly with the transition between major organic designs: vertebrates from invertebrates, for example, and jawed from jawless fish. No one has ever solved Mivart's old dilemma [Mivart was one of Darwin's critics] of "incipient stages of useful structures". For instance, the jaw is an engineer's delight; the same bones worked equally well to support the gill arch of a jawless ancestor. But can you really construct a graded series of workable intermediates? What good is a series of bones, detached from gills, but still too far back to function as a mouth? Did they move forward, millimetre by millimetre, finally to assume a co-ordinated position surrounding the mouth?'

Darwin was greatly troubled by problems of this type, and he added a complete chapter to later editions of *The Origin of Species* in an attempt to refute Mivart's criticisms. 'Basically, he answered that a graded series of intermediaries had to exist,' says Stephen Jay Gould, 'and our failure to specify their function merely expressed our lack of imagination.' Pinning his colours firmly to the mast of phyletic gradualism, Darwin said, 'If it could be demonstrated that any complex organ existed, which could not possibly have been formed by numerous, successive, slight modifications, my theory would absolutely break down.'

The opposite theory, which proposes evolutionary change through relatively rapid periods of modification, separated by long periods in which the species remains unchanged, does allow for *some* transitional forms, but it does not demand the existence of a long series of very finely graded intermediates as with phyletic gradualism. In this view of evolution, known as 'punctuated equilibrium', the anatomical change would be complete in ten or a hundred generations, and this transitional phase would be very short by comparison with the species' total duration. This explains the failure to find fossils of intermediate forms. Fossilization is, in the main, a rare event. The vast majority of animal remains are scattered and dispersed before they have a chance to become buried in deposits that will ensure fossilization. The chances of finding a transitional form of a species in the fossil record, when that transitional form represents a tiny fraction of one per cent of the total species population, are extremely small.

The origin of a new species, as envisaged by the punctuated equilibrium model, always takes place in a small group of individuals that are geographically isolated from the main population of the species. The new species arises there and later takes over the territory of the main species population, so that it would then appear in the fossil record in its fully developed form. According to this model, therefore, the incompleteness of the fossil record is accepted, inasmuch as fossilization is a rare event and there is never a complete record of year-by-year changes. The 'jerkiness' of the record is also accepted, however, as being a true reflection of the way evolution operates.

These, then are the two theories: a slow population-wide transition between one species and the next, or a brief evolutionary spurt separated by a long period without change. Does one have to choose between the two? As my good friend Alan Walker says, 'Evolution can surely work in both ways, and at all intermediate positions in between.' It is worth bearing this in mind when considering the woefully incomplete fossil record of human evolution. The immediate impression is that new grades of human ancestors suddenly appear in the record. There are, however, hints of intermediate forms too, as we shall see later.

The changing face of the earth

Apart from the cataclysmic changes brought about by earthquakes or massive volcanic eruptions, it is very easy to view the world in which we live as being a stable place. Mountains, valleys and rivers have the appearance of great age and permanence, but this is not so. Our life span is but a fleeting moment on the scale of geological time so that the continuous movements of the earth's crust are, to our eyes, imperceptible.

On the longest timescale of all is the steady drift of the world's continents. These immense landmasses ride on slowly moving tectonic plates that make up the earth's crust. Two hundred million years ago, all of the globe's landmass was in contact, forming a single supercontinent that we call Pangaea. With the steady movement of the tectonic plates, Pangaea split up, and the continents gradually travelled towards their present positions. Many features of the world as we see it today are relatively recent: North and South America were separate continents until about two million years ago and Africa was isolated from Europe by shallow seas until some time

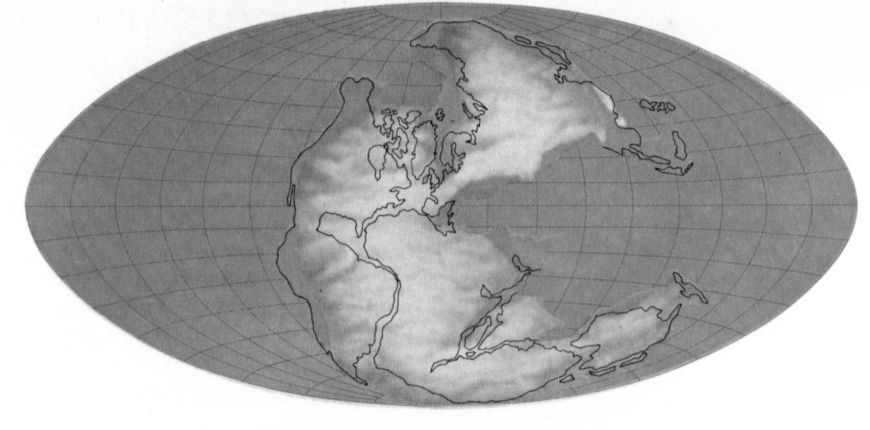

Two hundred million years ago all the continents were in contact, forming a single supercontinent known as Pangaea. The brown areas show dry land while the green areas represent shallow seas covering the continental shelves. The black lines indicate the position of the present-day continents in this landmass.

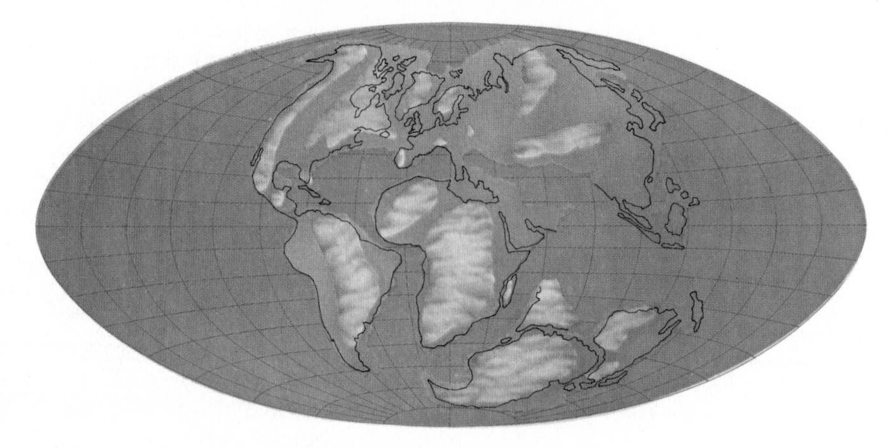

One hundred million years ago North America and Europe had begun to split apart, as had South America and Africa. The sea level was very high, so that there were many small islands of dry land separated by shallow seas.

between eighteen and sixteen million years ago.

Mountains are often formed when two tectonic plates are pushing against each other. The Andes, for instance, are being built up in response to the eastward movement of the Pacific plate, while the northward drift of the Indian subcontinent is crushing the edge of the Eurasian landmass, forming the massive Himalayan chain. It is here in the Himalayas that the tremendous forces of the earth's tectonic plates have led to the formation of one of the world's most important fossil sites.

David Pilbeam of Yale University is searching for early human ancestors in deposits that are about fourteen to eight million years old in the foothills of the Himalayas. As these mountains were steadily thrust upwards, erosion of the rising peaks also occurred, and, over millions of years, immense quantities of silt have been carried down in rivers and streams to be deposited on the lower slopes. There they buried the remains of the animals that existed at the time. The process of deposition went on for so long that the accumulating sediments reached an incredible 6 kilometres (3.7 miles) in thickness. In more recent times, the crumpling force of the moving Indian plate has begun to thrust these sedimentary deposits upwards, forming the Siwalik Hills. These hills, in their turn, are now being worn down by erosion and it is this process that is exposing the record of past ages, layered in the ancient deposits.

David's sites in the Siwalik Hills are a dramatic reminder of the restless-

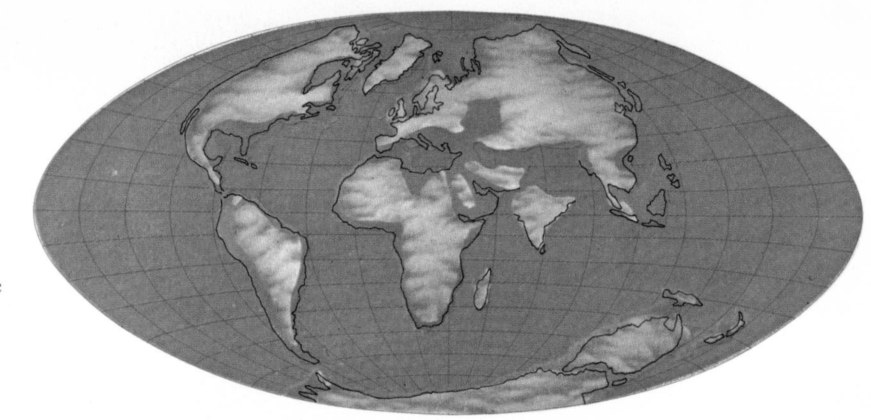

Fifty million years ago the Atlantic Ocean had begun to open up but South America and India were still islands, moving towards their present-day positions. Africa, too, was cut off from other landmasses because of the high sea level.

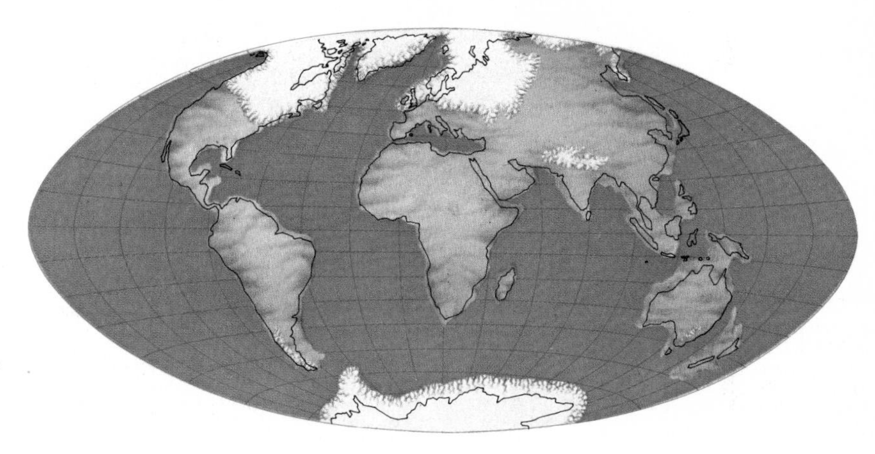

Forty thousand years ago the continents had almost reached their present positions. India had collided with the Eurasian landmass, forming the Himalayan Mountains. The last Ice Age was at its height and there was so much water locked up in the ice caps that the sea level fell, exposing most of the continental shelf areas.

ness of the earth's surface. Other sites are spectacular too, although formed by less dramatic forces. The Lower Omo Valley, at the point where the Omo River reaches Lake Turkana, is a site where fossils have been buried in deposits measuring up to a kilometre in thickness. All the Omo deposits have been carried in floodwaters from the Ethiopian highlands during the past four million years. During the same timespan, sediments were laid down by the rising and falling of ancient lakes to the north and south of the Omo River. In the Afar Triangle in northeast Ethiopia, 200 metres (650 feet) of sediments were deposited by a now vanished lake, and a joint Ethiopian, American and French team has been making spectacular discoveries in this area. Fossils and stone tools at Olduvai Gorge come from the deposits of an ancient lake which has now disappeared; the past eras are being exposed by a recent river valley cutting down through 100 metres (330 feet) of lake sediments.

When one walks on sedimentary rocks one is walking through time. At Olduvai Gorge, for instance, a scramble from the brim to the bedrock at the bottom takes you from sediments deposited just a few tens of thousands of years ago to fully two million years into our past, and yet the scramble is accomplished in a few minutes. At David Pilbeam's sites it is possible in some places to hike through fourteen million years in a day.

The accumulation of the products of erosion – silt and sand – followed by their transformation into sedimentary rocks, and then by further erosion, is

The Siwalik Hills, Pakistan. Erosion of the Himalayas brought down immense quantities of silt, which were deposited and formed sedimentary rocks. Further movement by the Indian tectonic plate then pushed these rocks up to form the Siwalik Hills. As these, in turn, are eroded, the fossils within them become exposed, yielding evidence about possible human ancestors of fourteen million to eight million years ago.

a common cycle in the changes of the earth's surface. It is also a crucial sequence of events as far as fossil-hunters are concerned. Bones must be covered quickly by sediments if they are to be preserved at all, but they are useless if they remain buried forever. Present-day movements of the earth's surface in areas such as the Himalayas and East Africa result in continuing erosion which is an essential ingredient in the search for evidence of our past.

The making of the fossil record

It is usually difficult to give anything more than a very rough estimate of how quickly sediments accumulated: was it over a period of ten years, one hundred years, or even several thousand years? One can tell whether the sediments were deposited from a river or at a lakeshore by microscopic examination of the rock, but the time taken to deposit each layer of sediment depends on many indeterminable factors. Breaks in the sequence of deposition, when a lake retreats temporarily or the rains fail so that rivers shrink or dry up, add to the problem.

Diane Gifford has observed periods of sediment accumulation on contemporary sites at Lake Turkana and has found that, at times, accumulation is very rapid. These observations have formed part of a study of how bones and stones enter the fossil record. At the end of September 1972, she had the opportunity of watching the fate of a butchery site occupied by a small group of Dassanetch hunters who killed a hippopotamus on the lakeshore. Because of the size of the carcase the men established a short-term camp about 60 metres (200 feet) inland from the kill. They carried most of the carcase to their camp, where they constructed two rock windbreaks and a stone-circled hearth. The group stayed there just a few days before moving on and leaving the debris of their camp-site, including the scattered bones of the hippo, behind them.

The lake was rising at the time, and very soon the head and neck of the carcase were standing in shallow water while the rest of the scattered skeleton was exposed on dry ground. The rise continued through to December of the following year, and then reversed. Finally the head and neck were again exposed above water, revealing a 3-centimetre (1-inch) layer of fine silt in and around the skull. This, together with the water, had served to preserve the bone. By contrast, the parts of the skeleton not covered by water had begun to crack and flake from exposure to the elements. This illustrates both the speed with which bones can be buried along a lake margin and the rapidity with which unprotected bone will begin to disintegrate.

At another camp-site Diane had a second opportunity to glimpse a sequence of events such as must have frequently happened with our ancestors in the past. Eight Dassanetch men and boys made a camp in a sandy streambed that was part of a delta system draining the flood plain. It was a comfortable place to camp as it was free from the potentially painful needle-points of the ubiquitous spike-grass. The site was 300 metres (1,000 feet) from the lakeshore.

The Dassanetch group occupied the camp for just a few days in November 1973, leaving behind them the remains of forty terrapins, four small crocodiles, fourteen catfish, a few Nile perch, other fish and some pieces of zebra carcase scavenged from lion kills. 'The rains in the following spring were unusually heavy,' Diane recalls, 'bringing down a lot of silt into the delta. Deposition from the first of the rains covered much of the site with 2 to 5 centimetres [1 to 2 inches] of silt and sand. This layer quickly dried, producing a hard cap over much of the site. Four days later much heavier rains came, and this time 5 to 10 centimetres [2 to 4 inches] of silt completely

Bone which is exposed to the sun and rain rapidly begins to crack and disintegrate, as can be seen in this elephant's skull. The other parts of the elephant's skeleton have been scattered by passing herds of animals, and many of the fragile, weathered bones have shattered under their hooves. Only bones which are rapidly buried by sediments are likely to become preserved and eventually fossilized.

covered the site. Later that summer, Kay Behrensmeyer and I excavated the site so as to examine the deposition and to discover what material from the camp still remained and what had been washed away. The results have allowed us to estimate the rates of water flow that selectively preserved and destroyed portions of the site.'

Events such as this must have happened many times during the history of the lakeshore because, through careful prospecting and excavation, we have discovered a number of ancient camp-sites all of which have a scatter of debris very reminiscent of that which Diane and Kay excavated. The discovery of the remains of the hominids who occupied these sites is, of course, vital to our attempt to build up a history of human origins. But the chances of any part of an animal skeleton finding its way into the fossil record are remote indeed.

One often comes across the victims of lion kills, mainly zebra and topi, near to the present lakeshore, and it is remarkable how rapidly the remains of these animals are dispersed. Hyaenas and jackals tear at the remaining

carcase, often dismembering it, and morsels are frequently taken away to be eaten in peace elsewhere. Exposure to sun and rain cracks the bones, and the trampling hooves of passing animals shatter them. Frequently the cranium and jaw bone are all that remain of a creature which died only a few months previously. The fate of hominid skeletons, whatever the cause of death, must have been very much like this in times past. It is no wonder that the search for hominid remains is so often fruitless. The occasions when a whole cranium is preserved, exposed, and discovered intact are rare indeed.

Major rivers draining into the Turkana basin formed the lake at least four million years ago, and its level has fluctuated ever since. At some time in its history, the waters of the lake flowed into the Nile, as the presence of Nile perch in today's closed lake testify. Around 10,000 years ago the water level was 100 metres (330 feet) higher than it is at present. Why it dropped so precipitously is something of a mystery, but several other East African lakes shrank dramatically at about the same time.

Each time the lake expanded it covered a greater and greater area, building up sediments as it went. Each time it shrank, areas that previously were underwater were drained and became subject to erosion. Major river systems also contributed to the growing layers of sediment, one of them being the Bakate River just south of the Koobi Fora spit on which our base camp is located. Bringing silt-laden water from the Ethiopian highlands, the Bakate River formed a large delta. This was an important depositional area and is rich in fossils. There is also a track of preserved footprints which were left by a human ancestor some one-and-a-half million years ago as he or she walked in the mud beside the lake. The tracks are now to be found in the ridges about a kilometre (half-a-mile) from the lakeshore.

The sediments of the ancient delta have been buckled by earth movements and this means that it is no longer a simple job to trace the layers of sediment from one area to another. The layers have also been jostled by faulting and tilting, and have been weathered by the seasonal streams which are brief but vigorous, creating a landscape of stark, hilly ridges and deep-sided gullies. Fortunately, however, at intervals through the deposits, there are layers of volcanic ash which are invaluable in helping us put dates on the fossils. These ash layers, or 'tuffs' as the geologists call them, either fell directly on the land after being spewed from nearby volcanoes or were brought down in rivers, to be dumped on the lake's flood plain. In either case, certain changes that occur in the mineral components of the volcanic ash during the heat and pressure of the eruption sets in motion an 'atomic clock'. This 'clock' can be used to date the tuffs: the method is known as potassium/argon dating. If a tuff is shown to be, say, 1.8 million years old, we know that anything buried below it is older than that date and anything above is younger. With luck another tuff will be lying just above at, say, 1.6 million years, which will give a very close estimate for the age of the fossils found in between the two tuffs.

The combination of good sedimentary conditions and the fact that animals, including hominids, like to be near a source of water, has made the eastern shore of Lake Turkana an extremely rich source of fossils. Although fragments of fossilized bone litter the sediments, the task of fossil-hunting is very skilled. 'It's easy if the fossil is white or grey,' says Kamoya Kimeu, 'because you can see it from 20 yards. But mostly the fossils are very dark and then it's difficult. After some practice you can see the fossils quite well, but you have to look very closely. If we find something interesting, we get on our hands and knees to be very near the ground. You can see very small pieces then.'

Kamoya takes his team out very early in the morning, usually at about

Part of the eastern shore of Lake Turkana, about two million years ago. Rivers draining from the Ethiopian highlands (shown in the distance) flowed into the Chew Bahir Lake which in turn supplied the Bakate River. As this river flowed into Lake Turkana, the silt it carried was deposited as a delta. Other layers of sediment were deposited from the lake waters themselves: the red layers represent these sediments.

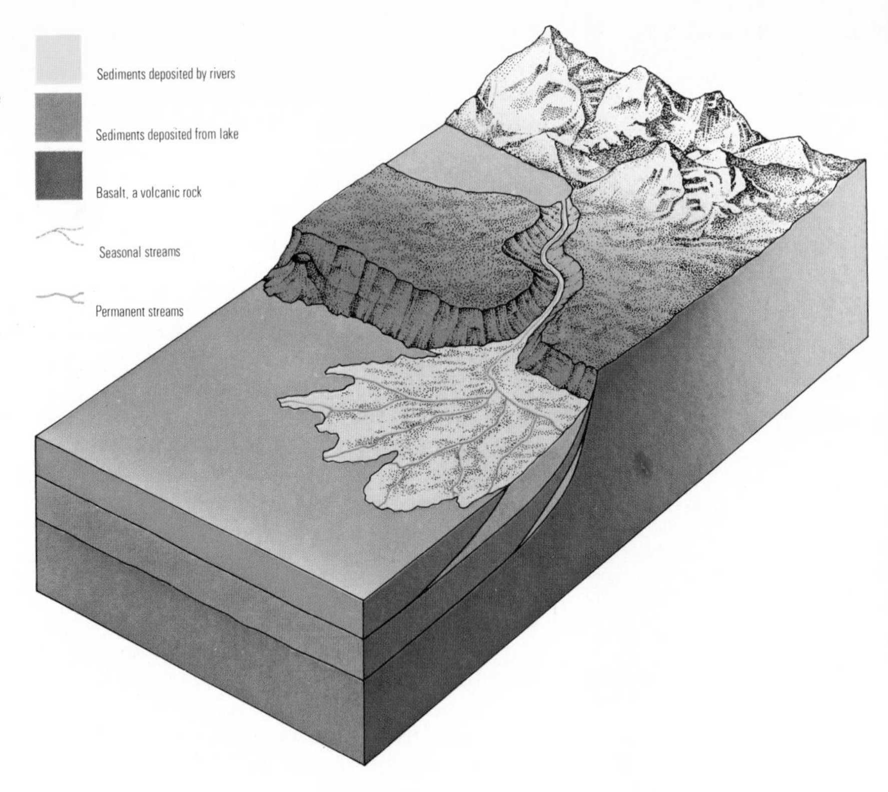

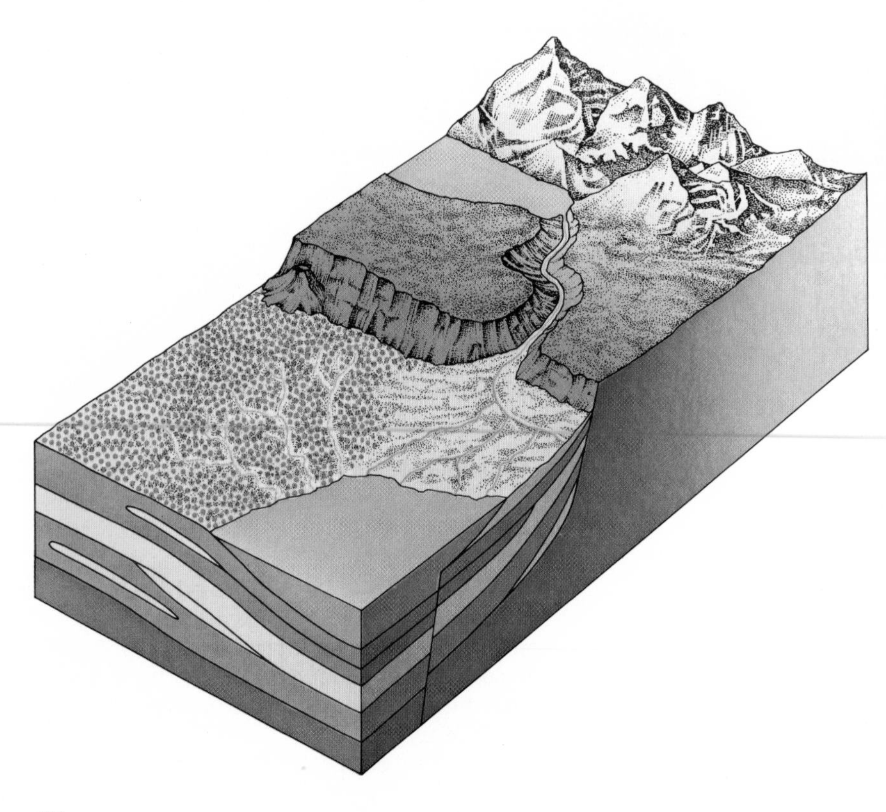

The same area about one million years later. The delta and lake sediments continued to build up. At times the lake level was high, while at others it fell, exposing large areas of the sediments. During these periods, seasonal streams flowed across the sediments, causing some erosion. Movements of the earth's crust associated with the formation of the East African Rift Valley caused the sediments to buckle and arch upwards. In addition, faults appeared in some places: these are shown by vertical lines.

The same section today. A drier climate has turned Chew Bahir into a salt lake and there is no flow of water through the Bakate Corridor at present. The sediments are being eroded by seasonal streams, which create a landscape of gulleys and ridges among the sedimentary rocks. The spit of sand on which the Koobi Fora base camp is sited can be seen in the foreground.

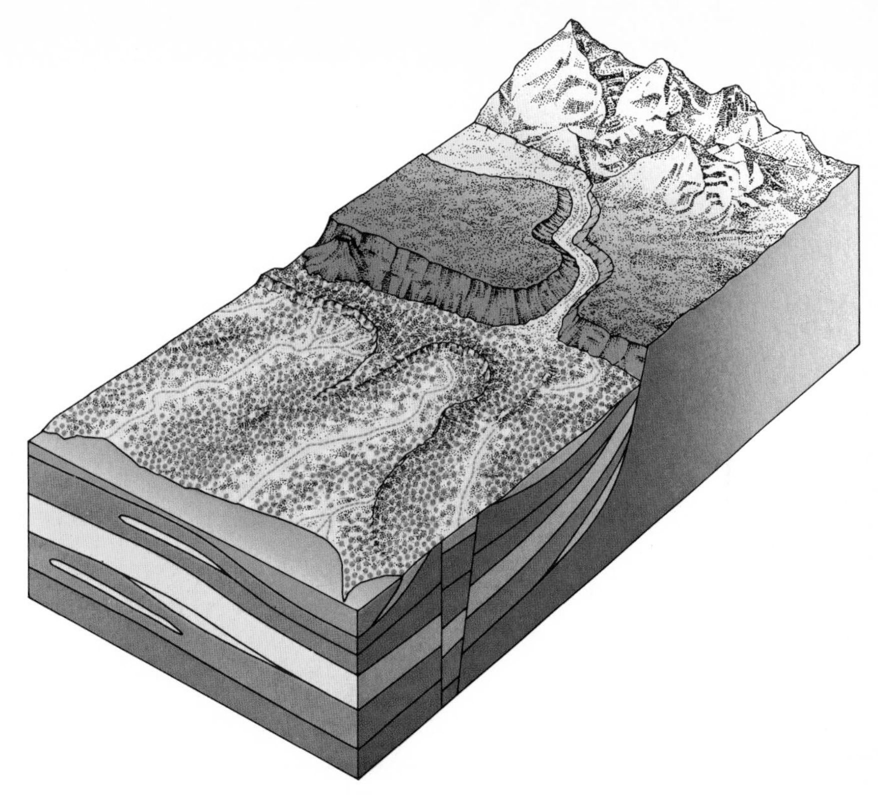

6.30 a.m., so as to work before the sun gets too hot. 'We drive to the area we want to search, and then walk in twos: there's always a danger of lions or snake-bites or something like that. And when someone finds something he can discuss it with his friend. It takes about three years before you can know the fossils very well. I know hominid bones very well now, but I don't always know what type it is.' To some of our visitors who are inexperienced in fossil-hunting, there is something almost magical in the way Kamoya or one of his team can walk up a slope that is apparently littered with nothing more than pebbles and pick up a small fragment of black, fossilized bone, announcing that it is, say, part of the upper forelimb of an antelope. It is not magic, but an invaluable accumulation of skill and knowledge.

3 Ape-like Ancestors

By an astonishing combination of circumstances, a very ordinary event which happened some three-and-three-quarter million years ago led to what is probably the most dramatic archaeological discovery of this century. Three hominids left a trail of footprints that have been clearly preserved, presenting us with an amazing picture of a few moments in the lives of some of our ancestors.

The place where it happened is now called Laetoli, a wooded area near a volcanic mountain, Sadiman, some 40 kilometres (25 miles) south of the present-day Olduvai Gorge in Tanzania. The dry season was probably nearing its end and the sight of gathering rain clouds promised welcome relief from months of drought. For a week or two the volcano had rumbled restlessly, occasionally belching out clouds of grey ash that settled over the surrounding countryside. Nothing violent or startling, just a steady background of subterranean stirrings such as is still experienced today from the Oldoinyo Lengai, 70 kilometres (45 miles) southeast of Olduvai. Like the ash from Lengai, the ash that Sadiman produced had a chemical composition that made it set like cement when dampened slightly and then dried by the sun.

As luck would have it, the end of that dry season was signalled by a scattering of brief showers. Large raindrops splashed onto the newly fallen ash, leaving tiny craters as on a miniature moonscape. The clouds passed; the promised downpour was yet to come. The carpet of ash was now in perfect condition for taking clear impressions: less rain and the ash would have blown about in the breeze, more rain and any impressions would have been washed away.

Following the rain, various animals left their tracks in the damp volcanic ash as they went their ways. Spring hares, guinea fowl, elephants, pigs, rhinoceroses, buffaloes, hyaenas, antelopes, a sabre-tooth tiger and dozens of baboons all made their marks. And so did three hominids. A large individual, probably a male, walked slowly towards the north. Following behind, then or a little later, was a smaller individual who for some reason placed his or her feet in the prints of the first individual. A youngster skipped along by their side, turning at one point to look to its left. The sun soon baked the prints into rock-hard impressions. More ash, rain and windblown sand covered and preserved the prints until they were discovered by a lucky chance in 1976.

'They are the most remarkable find I have made in my entire career,' says my mother, who is directing the excavations. 'When we first came across the hominid prints I must admit I was sceptical, but then it became clear that they could be nothing else. They are the earliest prints of man's ancestors, and they show us that hominids three-and-three-quarter million years ago walked upright with a free-striding gait, just as we do today.'

Mary Leakey examines the hominid footprints found at Laetoli, Tanzania.

A closer view of the hominid footprints. The craters made by the falling raindrops can also be seen, and a trail of animal prints leads off to the right.

Habitually walking around on the hindlimbs, leaving the forelimbs free for other jobs, is an unusual mode of locomotion. Once our ancestors had adopted an upright stance many things associated with being human became possible, such as fine manipulation with the hands, and the carrying of food back to a base camp. I do not mean to suggest that some four million years or so ago primitive hominids evolved upright walking in order to use their hands in refined ways or to develop a food-sharing economy. Indeed, this cannot be the case, because these behaviours did not arise until several million years after the development of upright walking. Nevertheless, the origin of bipedalism must be seen as one of the major steps, if not *the* major step, in human evolution. In this chapter I will explore the biological background and evolutionary pressures that produced a bipedal animal, and at the same time look at other evolutionary changes which occurred in early hominids, such as the modifications of the teeth.

The primate heritage

Homo sapiens belongs to the order of animals known as 'primates', which includes apes, monkeys, lemurs, pottos, bushbabies, lorises and tree shrews. Primates have a number of important characteristics in common, such as grasping fingers backed with nails rather than claws, and forward-pointing eyes giving binocular vision. These characteristics are of selective advantage to an animal living in the trees and preying on insects as the primate ancestor did. The earliest primate was probably a nocturnal animal about the size of a tree shrew. Later descendants adopted a daytime life of fruit-eating, an activity that demanded colour vision. The monkeys evolved some forty million years ago followed by the apes ten million years later. Last of all, some time between fourteen and four million years ago, came the hominids.

The seventy million years of primate evolution has been marked, among other things, by an increase in the size of the body. Running parallel with this there has been a striking advance in intelligence, and the primates have also developed an increasingly complex degree of social behaviour. However, perhaps the most notable characteristic of primates, particularly the higher primates, is opportunism. Survival depends primarily on maintaining a continuous supply of nutritious food, and the higher primates, perhaps best exemplified by baboons and chimpanzees, exploit every suitable food source available. The early hominids, especially the line that led eventually to modern humans, seem to have thrived because they extended this opportunistic behaviour. It should be emphasized that they did not do this by a conscious decision, but that the opportunistic streak evolved, as other biological traits do, because it was favoured by natural selection.

Of the primates, the chimpanzee is man's closest relative, while the two other great apes, the gorilla and orang-utan, are slightly more distant evolutionary cousins. The apes and hominids are collectively known as the 'hominoids'. Biologists would dearly like to know how modern apes, modern humans and the various ancestral hominids have evolved from a common ancestor. Unfortunately, the fossil record is somewhat incomplete as far as the hominids are concerned, and it is all but blank for the apes. The best we can hope for is that more fossils will be found over the next few years which will fill the present gaps in the evidence. The major gap, often referred to as the 'fossil void', is between eight and four million years ago.

David Pilbeam comments wryly, 'If you brought in a smart scientist from another discipline and showed him the meagre evidence we've got he'd surely say, "forget it; there isn't enough to go on".' Neither David nor others involved in the search for mankind can take this advice, of course, but we remain fully aware of the dangers of drawing conclusions from evidence that is so incomplete.

A tree shrew from Southeast Asia. The ancestral primate was probably a small, nocturnal animal, living in the trees and preying on insects, much as tree shrews do today.

The forest world of the dryopithecines

Fortunately, there is quite good evidence regarding the ape-like creatures that lived over fourteen million years ago. We know that about twenty million years ago Africa was the home of primitive, ape-like animals known as the dryopithecines. The globe was a good deal warmer then than it is now, possibly by as much as 12 °C (20 °F) in temperate latitudes. Thick forest carpeted much of the tropics and what is now Eurasia. The map of the globe was also different and Africa was still an island.

As far as one can infer from their skeletons and teeth, the dryopithecines were forest dwellers, feeding on fruit, soft leaves and shoots, flowers and probably insects. By contrast with today, when apes are few in number and all in danger of extinction, twenty million years ago they thrived. The apes were as numerous and successful then as monkeys are today.

There were many different species of dryopithecines and it is impossible to draw an accurate picture of how they were related to each other. One of these creatures was *Proconsul africanus*, a skull of which my mother discovered in 1948. Its teeth are ape-like in that the cheek teeth are small and have thin enamel, while the canines are sharp and projecting. *Proconsul* probably walked along branches on all four limbs rather than hanging below them in the manner of an orang-utan or gibbon.

Somewhere between eighteen million and sixteen million years ago Africa joined up with Eurasia, and various species passed from one continent to the other. This produced an explosion of evolutionary changes as animals were presented with new opportunities and new competition. New species arose while others were pushed into extinction. At about this time there is evidence that the world's climate began to cool, and this led to a steady shrinkage of the huge tropical forests. Thus the dryopithecines came under pressure from several directions.

Change, therefore, was inevitable, and a new group, the ramapithecines, appeared. These creatures were probably better adapted to the more open woodland environment that was spreading through much of the tropics and sub-tropics. Meanwhile the forest-dwelling dryopithecines diminished in number. It is to the ramapithecines that we look for signs of the first hominid ancestor, and possibly the ancestor of the modern apes too.

The move to open country

The ramapithecines can be distinguished from the dryopithecines principally by their teeth. The larger, flatter cheek teeth and generally smaller front teeth suggest that the ramapithecines were having to deal with tougher and less nutritious food items than is normal for apes. It is from this change in the teeth that palaeontologists infer a shift from the forest to more open woodland. David Pilbeam explains: 'Food items in forests are usually soft, such as fruit and young shoots, whereas in woodland they are not only tougher, such as roots, but they are also less nutritious. The ramapithecines therefore probably had to process more food and process it more thoroughly too.' At roughly the same time as the ramapithecines emerged, the fossil record shows that other mammals, such as antelopes, pigs, elephants and rodents, were evolving adaptations to the newly available open-country environment.

The ramapithecines flourished between fourteen and eight million years ago and they were spread widely across Africa, Asia and Europe. One of the best areas for finding ramapithecine fossils is the Siwalik Hills, where David Pilbeam and Ibrahim Shah of the Geological Society of Pakistan have been conducting a joint project since 1973. They are doing more in the Siwaliks than simply making a collection of fossils. 'The idea,' says David,

The skull of an African dryopithecine which was discovered by Mary Leakey in 1948, and given the name *Proconsul africanus*. The ape-like dryopithecines were a very successful and varied group, but with the shrinking of the world's forests about sixteen million years ago they suffered a decline in numbers.

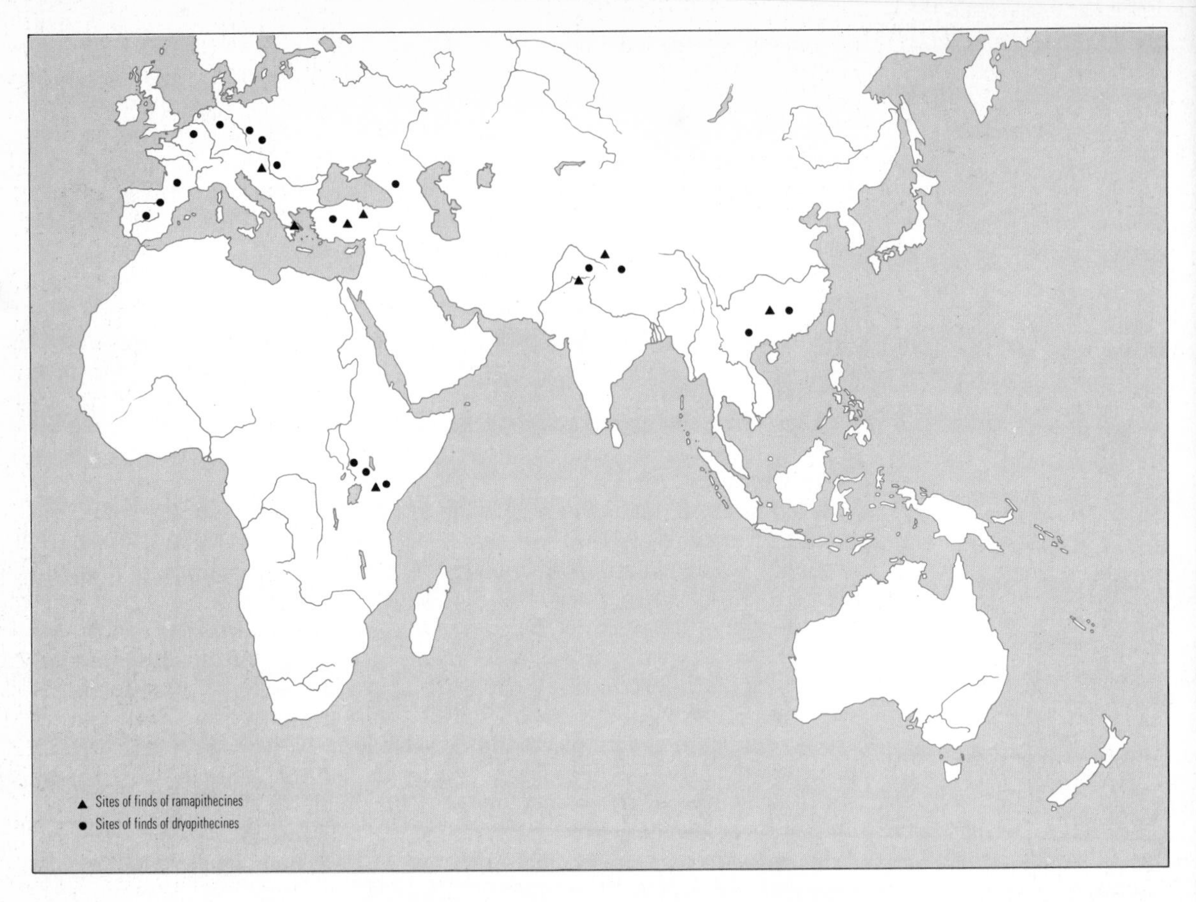

Sites of finds of dryopithecine and ramapithecine fossils.

'is to build up a picture of the environment in which these creatures lived, to see how it changed through time and to determine how this affected the animals living there.'

Martin Pickford, a member of the team, describes the environment in this way: 'There was no such thing as an "average" habitat. Some areas were clearly pretty thickly covered by trees: browsing animals such as pigs, giraffes and small antelopes lived there. Other places must have been relatively open, with animals such as horses and large antelopes. In still other locations we find remains of turtles, crocodiles and hippo-like creatures, indicating an aquatic environment. It was very much a contrasting mosaic.' It is in the partially wooded areas that the ramapithecine fossils are found.

The group has been extremely successful in retrieving fossils from these extensive deposits, but unfortunately all the specimens are rather fragmentary. Nevertheless, a pattern is emerging, as David Pilbeam describes: 'We find three sizes of hominoids, with three sizes of skulls and jaws, and other parts of the body. As a reasonable guess, I've associated the largest of the skull fragments with the largest pieces from the rest of the skeleton, and so on. The smallest of the three is *Ramapithecus*, after which the group is named. When alive, it probably weighed around 20 kilograms (45 pounds) and probably split its time between life in the trees and the ground. The next is *Sivapithecus* which was very similar to *Ramapithecus*, but somewhat bigger. It too was probably half arboreal and half terrestrial. The last was *Gigantopithecus* which, as its name implies, was a hefty creature. Indeed, one late species from China reached truly gigantic proportions, and has been

Opposite: Three types of ramapithecines can be identified among the fossils from the Siwalik Hills: this specimen belongs to the group named *Sivapithecus*.

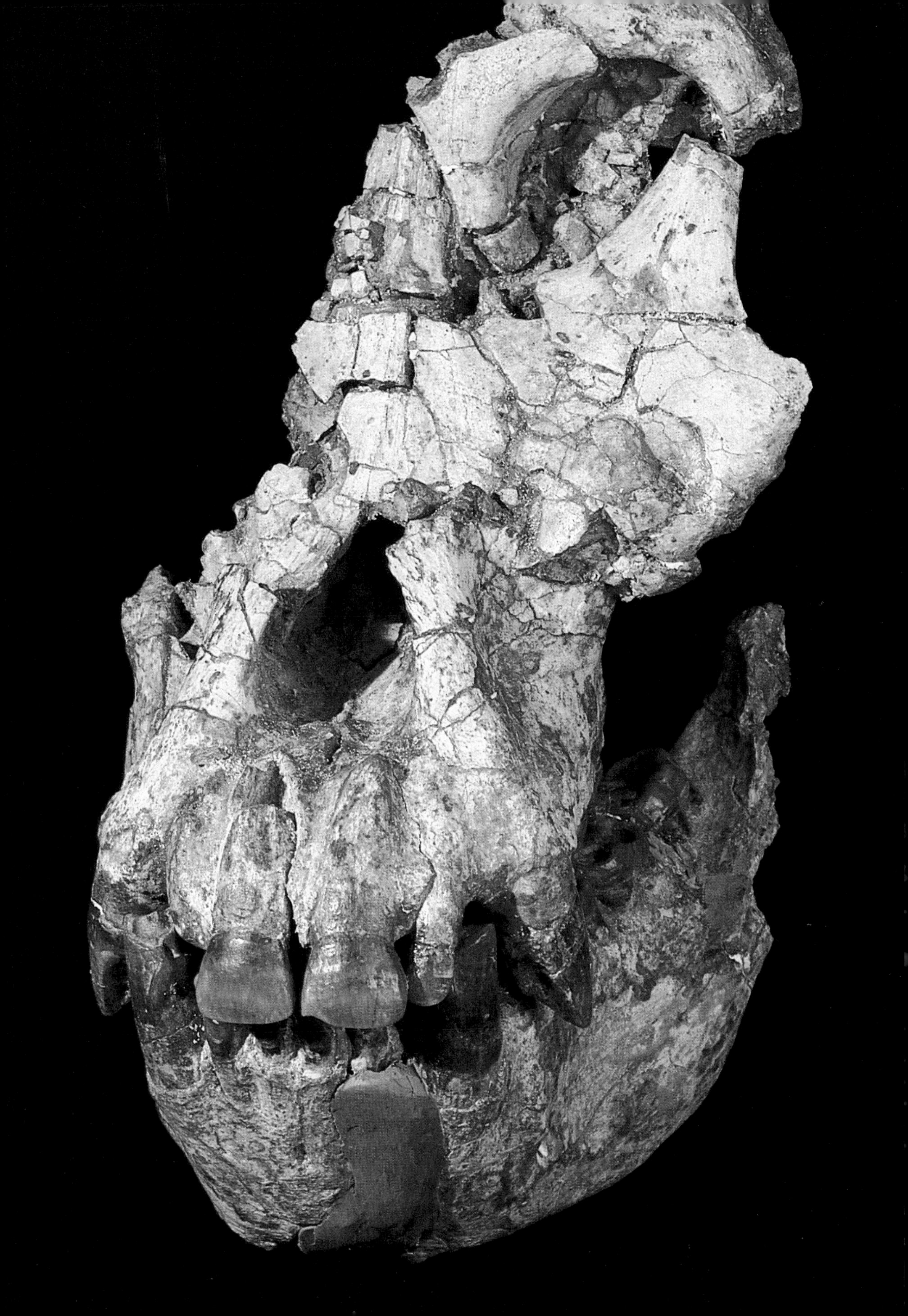

associated with mythical stories of the Yeti and so on. Because of its size, *Gigantopithecus* probably spent most of its time on the ground.'

Although the layers of time encapsulated in the Siwalik deposits cover a period from fourteen million years to a mere half-million years ago, the fossil hominoids are found only in the first half of the sequence. After what appears to have been a brief blossoming of hominoid populations around eight million years ago, these primates vanished from the Siwaliks. 'It seems to me that a lot of changes were going on around six to eight million years ago,' suggests David. 'There's evidence from ocean deposits that major changes in currents were associated with a lurch to a cooler climate. Perhaps this is the reason the hominoids could no longer inhabit Pakistan. It's during this period too that many of the modern plains animals of Africa arise. All the major groups of open-country animals really get going at this time. There's a lot of evolutionary activity going on, probably in response to environmental change.'

Although the ramapithecines disappeared from Pakistan as the climate cooled, they may well have continued to inhabit Africa. They had been on the African continent for a long time, and one of the oldest known *Ramapithecus* fossils was found by my father at Fort Ternan in Kenya. During the period when *Ramapithecus* flourished, the African continent was undergoing important changes. There was an updoming of the earth's crust and highland areas were formed which had not existed before. The new mountain ranges brought new environments and a change in climate.

In addition to this, the African continent began to pull apart slowly, and faults formed in the rock, resulting in the East African Rift Valley. The valley runs from the southernmost tip of Turkey, down through Israel, along the Red Sea and into Africa as far as the mouth of the Zambezi river. It varies in width and depth, but in places it is 80 kilometres (50 miles) wide and over 300 metres (1,000 feet) deep. The Rift Valley produced dramatic changes in altitude and drainage over very short distances and this, together with the newly formed mountain ranges, resulted in a far more complex and varied mosaic of habitats. Such a diversity of environments provides an ideal setting for evolutionary change, and the new and varied opportunities were undoubtedly exploited by the expanding hominoid populations. It is possible that the rich array of environments generated by the formation of the Rift Valley and the new mountain ranges were instrumental in the evolutionary origin of the first upright hominoid: in other words, the first hominid.

During the past decade, *Ramapithecus* has been widely considered as a candidate for 'the first hominid', which would put the beginning of the human line and the split from the apes at around fourteen million years ago. The principal reason for considering *Ramapithecus* as the ancestral hominid is that its teeth are very similar to those of later hominids. It is by no means certain, however, that *Ramapithecus* survived from fourteen million years ago to some time within the 'fossil void' when it could have evolved to give rise to primitive hominids. It is just as likely that *Ramapithecus* eventually died out, and the first hominid arose from an ancestor yet to be discovered. As David Pilbeam says, 'It's best to keep an open mind on these things for the moment. *Ramapithecus*, or something derived from it, *may* be the first hominid, and *Sivapithecus* is a reasonable model from which to derive the orang-utan, but we simply can't be sure.'

Opposite: The East African Rift Valley was formed as the earth's crust pulled apart, forming cracks or faults. The formation of the Rift Valley may have created a greater diversity of environments in which *Ramapithecus* could have evolved to give rise to primitive hominids. Many prehistoric sites are associated with the Rift Valley because of the lakes around which animals, including hominids, would tend to gather, and where ideal conditions existed for both the preservation and subsequent discovery of their remains. The major African sites mentioned in this book are marked on the map.

Molecular evidence

During recent years, Vincent Sarich and Allan Wilson, two biochemists at the University of California, have been studying the differences between

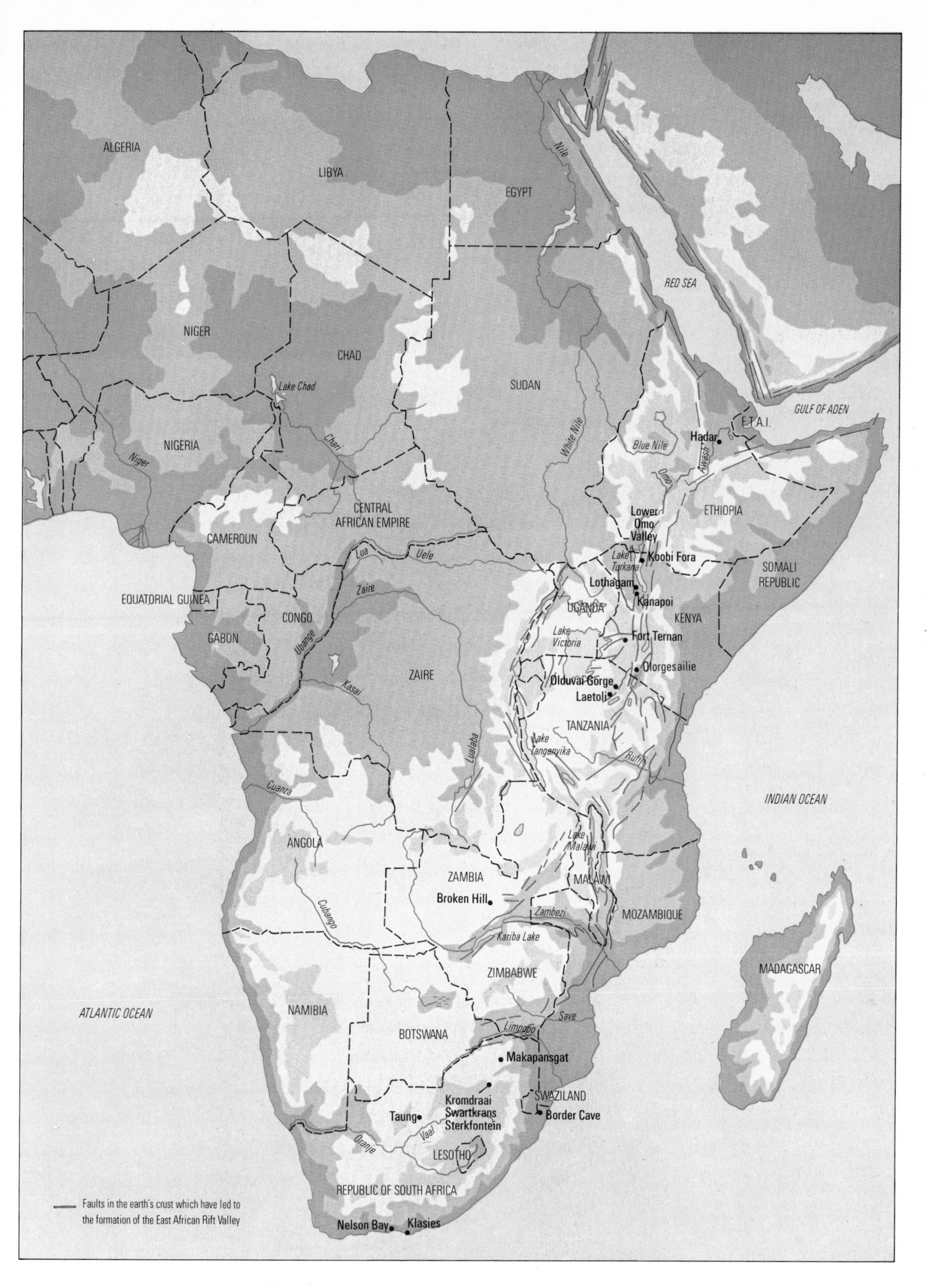

ALGERIA
LIBYA
EGYPT
Nile
RED SEA
NIGER
CHAD
SUDAN
Lake Chad
GULF OF ADEN
F.T.A.I.
NIGERIA
Niger
Chari
White Nile
Blue Nile
Hadar
Awash
Omo
ETHIOPIA
CENTRAL
AFRICAN EMPIRE
Lower
Omo
Valley
CAMEROUN
Lua
Uele
Lake
Turkana
Koobi Fora
SOMALI
REPUBLIC
Lothagam
Kanapoi
EQUATORIAL GUINEA
Zaire
UGANDA
KENYA
CONGO
GABON
Ubange
Lake
Victoria
Fort Ternan
Olorgesailie
ZAIRE
Kasai
Olduvai Gorge
Laetoli
TANZANIA
Lualaba
Lake
Tanganyika
Rufiji
Cuanza
INDIAN OCEAN
ANGOLA
Lake
Malawi
ZAMBIA
MALAWI
Broken Hill
Zambezi
MOZAMBIQUE
Cubango
Kariba Lake
MADAGASCAR
ZIMBABWE
ATLANTIC OCEAN
NAMIBIA
Save
BOTSWANA
Limpopo
Makapansgat
SWAZILAND
Kromdraai
Swartkrans
Sterkfontein
Taung
Border Cave
Vaal
Oranje
LESOTHO
REPUBLIC OF SOUTH AFRICA
Faults in the earth's crust which have led to the formation of the East African Rift Valley
Nelson Bay
Klasies

Ramapithecus, the smallest of the ramapithecines, is considered the most likely candidate for the ancestor of the hominids. However, the remains of *Ramapithecus* that have been found are very meagre: only a few teeth and jaw bones. A great deal can be deduced from these finds by experienced palaeontologists, but much must remain speculative. The photograph on the left shows one of the best preserved *Ramapithecus* specimens found, and the drawings show the variety of reconstructions that have been made on the basis of such finds.

human and chimpanzee proteins. A protein is made up of long strings of amino acids, and changes can arise in these amino acids as a result of gene mutations. Some of the amino acids are vital to the protein's structure and function, but many are not, and these can change without affecting the way the protein acts in the body. Such changes are 'invisible' to natural selection, so they are not eliminated. Thus they tend to accumulate as time goes by. The rate at which mutations occur is fairly regular and so the number of differences between the proteins of two animal species can be used as an approximate measure of how much time has elapsed since they split off from a common ancestor and began to evolve separately.

The data which Vincent Sarich and Allan Wilson have collected on human and chimpanzee proteins suggest that we began to diverge only five million years ago. This figure is astounding to palaeontologists and would certainly rule out the fourteen-million-year-old *Ramapithecus* as a common ancestor. However, it is known that this biochemical method of investigating evolutionary relationships and the timing of evolutionary events is not always completely reliable: the 'molecular clock' can sometimes run too fast or too slowly. The issue remains to be resolved but this study of proteins has been beneficial in many ways, particularly in forcing researchers to re-examine their assumptions about man's early evolution.

The origin of upright walking

The principal characteristics of the earliest hominids were probably a change in the dental apparatus, such as is seen in the ramapithecines, and the adoption of an upright posture and gait. Why did these new features arise? It is worth stressing once again that the changes did not arise as part of an inevitable trend towards modern man: evolution does not work in a purposeful or directed manner. As already explained, the dental changes seen in *Ramapithecus* may well have been an evolutionary response to the spread of more open terrain. But why did our ancestors take to upright walking, and at what stage did this occur? Over the years there have been many different explanations.

The freeing of the hands for the making and use of tools and weapons has been a popular explanation of bipedalism for a very long time. But in fact there is no sign of any artifact in the archaeological record until around two million years ago, which is at least two million years after hominids adopted upright walking. 'I used to think that cultural behaviour and toolmaking were the important factors in the split between the hominids and the modern apes,' says David Pilbeam, 'but I don't see it that way any more. Some biologists are beginning to look away from the head and towards the stomach for explanations of what went on in human evolution. Food, the way it's collected and processed, is vital in determining animal behaviour. And by taking this approach, anthropology is moving much closer to the study of other mammals in the search for hominid origins.'

The first major attempt to understand human origins in terms of feeding came just a decade ago when Clifford Jolly of New York University proposed his 'seed-eating hypothesis'. 'Look at gelada baboons,' he suggested, 'they spend much of their feeding time squatting in a fairly upright position while picking up small food items from the ground. They have small canine teeth which allows the jaws to swing freely in grinding their food. Hominids were probably living in open country where grass seeds and other small items would be abundant. Their dental features were well suited to dealing with tough seeds and so on.' As Cliff Jolly would himself now admit, this is probably only part of the story and it is not a particularly good explanation of bipedalism. Cliff Jolly's 1970 paper was, however, a milestone in the study of human prehistory in that it established an important new approach to the study of human origins.

How else, then, might one explain bipedalism in terms of diet? In an open woodland habitat, the food items, as well as being tougher, also tend to be more scattered or widespread than in the forest. An animal living in woodland or bush savanna spends most of its time on the ground while travelling from one food source to the next. As David Pilbeam points out, 'The contexts in which living primates do stand or walk upright are for the most part when they are feeding and on the ground.' A crucial issue is the size of the body, in that a large primate would experience more problems in adapting to an upright gait. A small ape, such as a gibbon or siamang, can manage to walk upright with much more ease than a gorilla or chimpanzee. 'I believe it is quite reasonable to conclude that an animal as small as *Ramapithecus* would have walked upright very readily when it was on the ground,' David suggests. 'And this habit would have been reinforced if it spent a good deal of time feeding while on the ground, taking fruit, berries and nuts from low bushes, for instance.'

David emphasizes the point that the end product of an evolutionary change may be totally unrelated to its original adaptation: 'Maybe bipedalism is analogous to the cheek pouch in certain monkeys: it enables you to carry food away from other individuals so you can eat it in peace. So, even though bipedalism eventually allowed the development of the highly social activity of the hunting-and-gathering way of life, the initial evolutionary pressure may have been to avoid sociality.'

Whatever the reason for bipedalism, the switch from being a four-footed creature to being bipedal is dramatic. Owen Lovejoy of Kent State University has made a close study of bipedalism and describes the transition in these terms: 'It's an absolutely enormous anatomical change. There are important changes in the bones, the arrangement of the muscles that power them, and the movement of the limbs. There are modifications to the internal organs too, but these are not as radical. Overall, the move to bipedalism is one of the most striking shifts in anatomy you can see in evolutionary biology.'

The split between apes and hominids

David Pilbeam has been the leading figure in recent years in assessing early hominoid fossils, so it is appropriate to give his current guess (and he would be the first to admit that it is a guess) of the 'timetable' of hominoid evolution: 'I would say that the last common ancestor between all the hominoids – that is the Asian apes (orang-utan, gibbon and siamang), the African apes (chimpanzee, pygmy chimpanzee and gorilla), and the hominids – was around ten million years ago. The Asian apes then split off. The division between the African apes and the hominids then took place roughly seven million years ago, at the time of extensive climatic and environmental changes. But only by retrieving a great deal more good fossil evidence will schemes such as this be testable in any really scientific way.'

Facts have often been in short supply in the study of human prehistory, and David describes very neatly the way that ideology has exerted a substantial influence on people's interpretation of the past: 'In Darwin's day, when evolution and life was seen as a battle, he emphasized the use of tools as weapons. In the early decades of this century – the heyday of Edwardian optimism – the brain, intelligence and higher thoughts were said to be what made us what we are. The obsession with the brain is precisely why the Piltdown forgery, which had a modern-sized cranium associated with an ape's jaw, was accepted as genuine with such alacrity. In the 1940s, with the burgeoning of technology, Man the Tool-maker held the stage. The war years left their mark with the rise of man being linked with an ancestry as a "killer-ape". And it's surely no accident that the blossoming of the media in the 1960s coincided with a turn to language as the engine of human advance. Now, with the strength of the women's movement growing, the role of the male in Man the Hunter is being replaced by a picture of co-operative hunting-and-gathering groups in which females play a leading role.' When considering our origins it is clear that we have often been less than objective.

Science is often seen as a search for answers that, given sufficient time, must surely be forthcoming, but, because of the nature of the evidence – or rather, lack of it – this may not be the case in palaeoanthropology. David Pilbeam concludes, 'It's my conviction that there may be many aspects of human evolution that will always elude us. We should be straightforward and honest about that. So far the spotlight has always swung to those who appear to come up with answers. It's salutary to consider that perhaps in future the prizes should go to those people who are able to differentiate between those questions that have answers and those that do not.'

Why did our ancestors take to upright walking? David Pilbeam believes that bipedalism may have allowed our ancestors to carry food away and eat it in peace, as this young baboon is doing. Having discovered a cache of ostrich eggs, it is seen here carrying off one egg before the rest of the baboon troop arrives on the scene.

4 The Early Hominids

In the previous chapter we saw *Ramapithecus*, who may have been ancestral to the hominids, venturing into far more open country than its predecessors and, perhaps, beginning to walk upright. *Ramapithecus* disappears from the fossil record eight million years ago and there follows a period of four million years which remains obstinately blank, in spite of the efforts of many palaeontologists. Then, from about four million years ago onwards, several hominid species begin to appear, and we first see creatures that can certainly be called our ancestors.

David Pilbeam and Stephen Jay Gould once opened a serious scientific paper with the following tongue-in-cheek statement: 'Human palaeontology shares a peculiar trait with such disparate subjects as theology and extra-terrestrial biology: it contains more practitioners than objects for study.' Fortunately, prehistorians are now in a much stronger position, following a decade of unprecedented discoveries. The 1970s were punctuated with newspaper headlines announcing 'yet another dramatic find'. At times the pace of discovery has been breathtaking, and since the new fossils were followed by a rethinking of human prehistory, the pages of our past have been almost entirely rewritten during the last ten years. Admittedly, there has yet to be proposed a coherent theory with which everyone whole-heartedly agrees: that would be too much to expect in a fast-moving and often emotionally charged science.

Nevertheless, there is a growing cohesion in the attempts to trace the course of human history. This has a lot to do with the marrying together of a number of different scientific disciplines, such as geology, taphonomy (the study of how a bone becomes a fossil), ecology, primatology and molecular biology, with the more traditional pursuits of archaeology and palaeontology. Furthermore, it is now possible to ask more pertinent questions than was possible hitherto. We can ask, for instance, how the basic mechanisms of evolution operated in the early stages of the emergence of mankind; how our ancestors made a living; what their ecological relationship with their close evolutionary relatives was; and what behavioural features separated those hominids that eventually became fully human from the ones that disappeared.

The Piltdown forgery

One difficulty with the search for man's place in nature is that the desire to 'know' has often been accompanied by a firm preconception of what would be found. So it was in the early years of this century, when amateur archaeologist Charles Dawson 'discovered' the Piltdown skull, an ape-like jaw associated with a large modern-looking cranium. The discovery, made in 1912 at a gravel pit in southern England, fitted neatly the prevailing view of a human ancestor as having well developed intellectual powers but

Swartkrans, one of four cave sites in the South African Transvaal which have yielded hundreds of fossils of early hominids. Bob Brain (far left) has been excavating at Swartkrans for fifteen years, and has contributed greatly to our understanding of how these cave deposits accumulated.

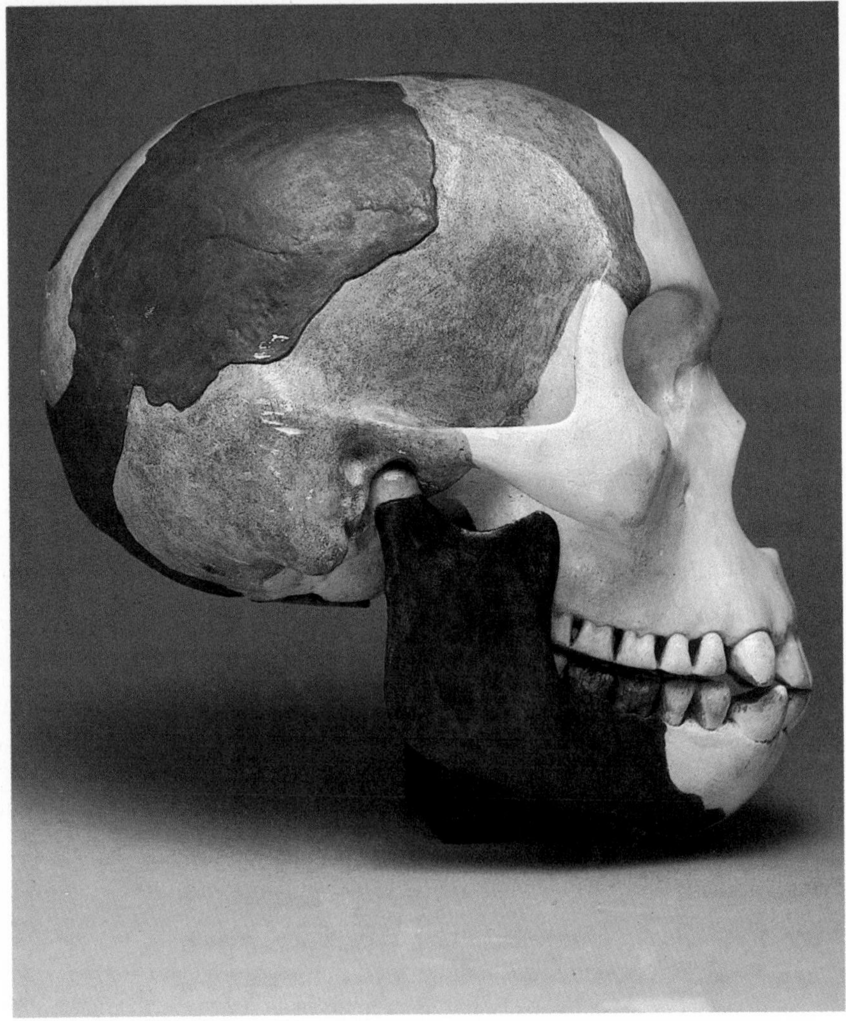

Right: The Piltdown skull, 'discovered' in a gravel pit in southern England in 1912. Guarded from scientific scrutiny for several decades, it was only revealed as a forgery in 1955.

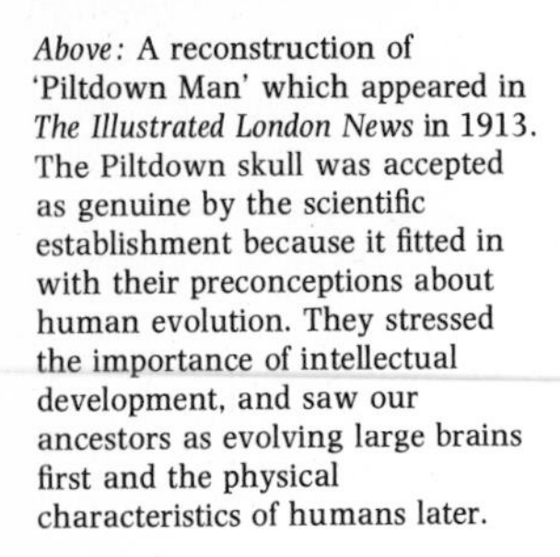

Above: A reconstruction of 'Piltdown Man' which appeared in *The Illustrated London News* in 1913. The Piltdown skull was accepted as genuine by the scientific establishment because it fitted in with their preconceptions about human evolution. They stressed the importance of intellectual development, and saw our ancestors as evolving large brains first and the physical characteristics of humans later.

some ape-like physical features. It also delighted British prehistorians, as the discovery swung the scientific spotlight away from the continent of Europe and on to the centre of the British Empire: the first man was clearly intelligent . . . and an Englishman!

A number of scientists were sceptical of the validity of the Piltdown skull, including my father, who asked permission to study the original specimens in 1933. Like many such requests, Louis's was refused and, after only a brief look, the material was returned to a safe, and he was just left with casts. It turned out in 1955 that the skull was a blatant forgery. Someone – there is still hot debate as to who it was – had put together a human cranium and an orang-utan's jaw, both suitably treated to give them a patina of age. The scientific establishment completely fell for the trick, not because the Piltdown skull was demonstrably old and genuine, but because it matched powerful preconceptions of what our forebears were like.

It was against this background that Raymond Dart announced in 1925 his discovery of the Taung child, an infant 'man-ape' from a limestone cave in the Transvaal, South Africa. British prehistorians immediately dismissed the claim. Sir Arthur Keith wrote in scathing terms, saying that the specimen was 'the same group or sub-family as the chimpanzee and gorilla . . .'

The Taung child, the skull of a young australopithecine found by Raymond Dart in 1924. He announced the discovery the following year, but it was dismissed by most scientists as being nothing more than an extinct type of ape. Further discoveries confirmed that these creatures were indeed hominids, though not our direct ancestors.

and that 'to make a claim for the Taung ape as a human ancestor is therefore preposterous.'

The fossils of the South African caves

The mistake, however, was with the scientific establishment, not with Raymond Dart. His proposition had to wait nearly twelve years before it received material support, in the form of an adult man-ape discovered in another limestone cave, Sterkfontein, by Robert Broom. Slowly, very slowly, the scientific tide began to turn, and Raymond Dart's and Robert Broom's discoveries were acknowledged as opening a new era in the search for human ancestors: they had found *Australopithecus africanus*, a hominid that lived in Africa between three million and one million years ago. The word *Australopithecus* simply means 'southern ape', so this Latin name for the first early African hominid to be found says, uncontroversially, 'southern ape of Africa'.

In June 1938, a second type of man-ape turned up. Robert Broom recognized that a number of teeth and several parts of a cranium from another limestone cave, Kromdraai, belonged to a stockier form of man-ape than the one he had found at the Sterkfontein cave. The new hominid came to be known as *Australopithecus robustus*, the 'robust southern ape of Africa', a name that reflects its large cheek teeth and the prominent bony attachments for its jaw muscles.

Raymond Dart demonstrated exceptional powers of perception in recognizing the special characteristics of the diminutive Taung skull at a time when the prevailing scientific opinion was certain to be hostile. And Robert Broom's enormous energy, drive and enthusiasm in pursuing the initial discoveries left onlookers gasping with awe and admiration – he was already seventy years of age when he propelled himself into his new career as a human palaeontologist. Indeed, both men were legends in their own time. All of us involved in the search for human origins owe them a great debt.

For more than a quarter of a century South Africa was the focus of man-ape discoveries, with the main findings coming from the Sterkfontein, Kromdraai, Swartkrans and Makapansgat caves. The first three of these lie in a tight group halfway between Johannesburg and Pretoria, while Makapansgat is some 250 kilometres (150 miles) northeast of Pretoria. The Transvaal is widely covered with limestone, and water seepage over long periods of time has carved out extensive cave systems that often open to the surface through vertical sinkholes. These rich deposits of lime attracted quarrymen to the area. Indeed, all the caves I have mentioned have had a dual role as lime quarry and archaeological site. As Alun Hughes, Raymond Dart's long-time colleague, puts it: 'We have to thank the quarrymen because they opened a lot of areas up to us, but we curse them too, for the fossils they must have destroyed.' So extensive has the quarrying been at Taung that it no longer exists as a prehistoric archaeological site.

The caves are important to prehistorians because they are places where, by some means, bones collected in enormous numbers: from Makapansgat, for instance, Raymond Dart and his colleagues have collected a quarter-of-a-million animal bones. Such glimpses of the past are potentially of tremendous importance to prehistorians, but there are problems. Firstly, there is a practical difficulty in retrieving the bones, and secondly, the problem of ascertaining how and when they came to rest in the cave.

Every time it rains on the high veld, water pours into the caves carrying with it mud, leaves and anything else that is in its path. Gradually the underground caverns fill up with deposits, trapping any bones that are there between slowly solidifying layers of silt. If the process of cave-filling had been as straightforward as this, there would still be problems enough for the archaeologist and prehistorian, but as Bob Brain, director of the Transvaal Museum, has found to his considerable cost, the story is even more complicated.

Bob Brain has been excavating at Swartkrans for fifteen years, often working through the rubble left behind by the quarrymen. The cave contents are thought to be just less than two million years old and accumulated on at least two separate occasions. Until recently Bob Brain had the impression that the periods of infill went on for a long time, perhaps a million years, but now he believes that the deposition was a short-lived process, going on for 10,000 years or even less. The two separate layers of deposits in the cave, therefore, seem to represent two discrete windows into the past, the older one underneath, the younger on top. Their exact age is difficult to determine, but one can make guesses based on the evolutionary stage of the animal bones in the deposits.

Dating problems aside, Bob Brain has recently discovered to his dismay that there is a further complication. It appears that after the first layer was

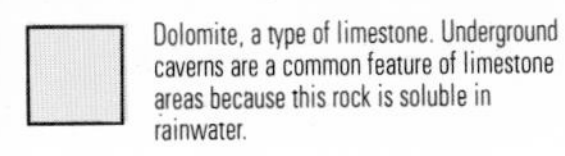

Dolomite, a type of limestone. Underground caverns are a common feature of limestone areas because this rock is soluble in rainwater.

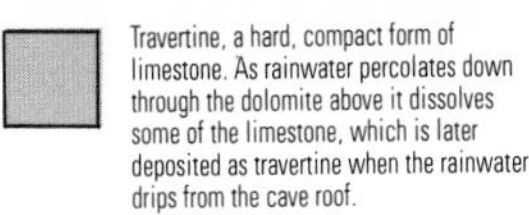

Travertine, a hard, compact form of limestone. As rainwater percolates down through the dolomite above it dissolves some of the limestone, which is later deposited as travertine when the rainwater drips from the cave roof.

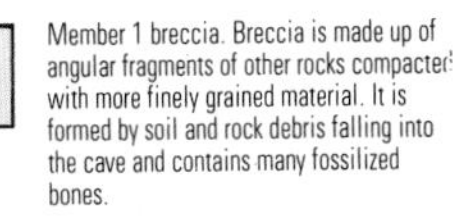

Member 1 breccia. Breccia is made up of angular fragments of other rocks compacted with more finely grained material. It is formed by soil and rock debris falling into the cave and contains many fossilized bones.

Member 2 breccia. Although formed in the same way as the member 1 breccia, this younger breccia is brown rather than pink. Certain chemicals give the rock its brown colour and in the older breccia these have been leached out by rainwater.

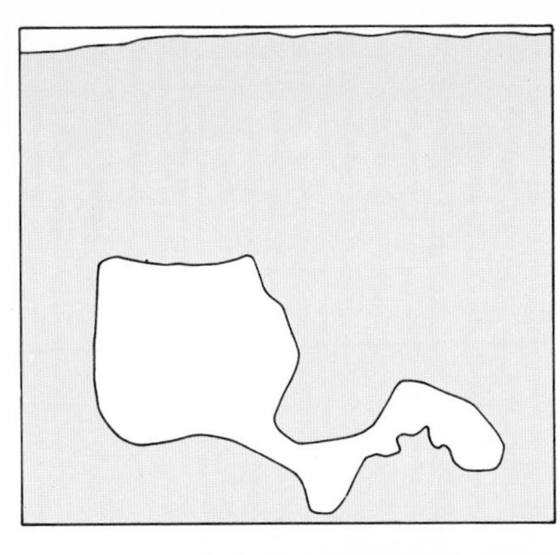

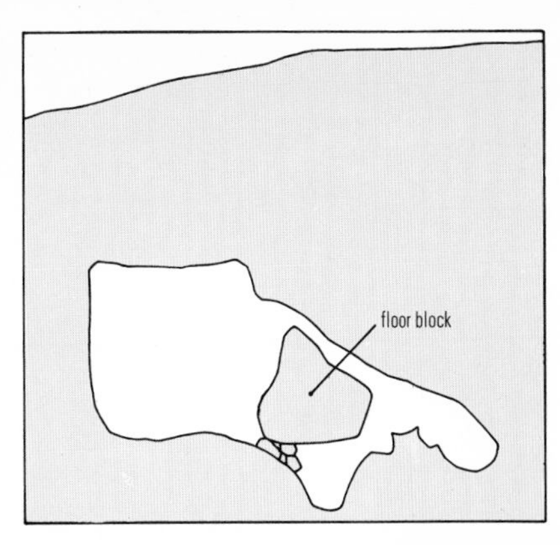

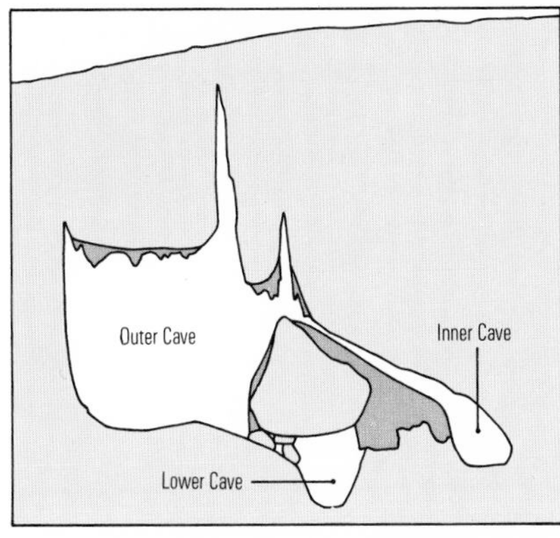

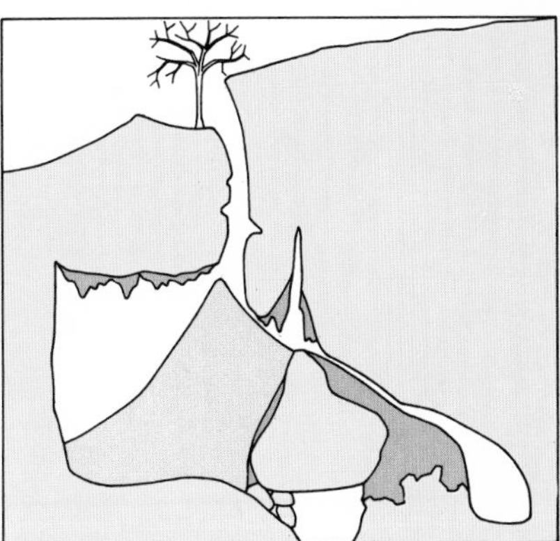

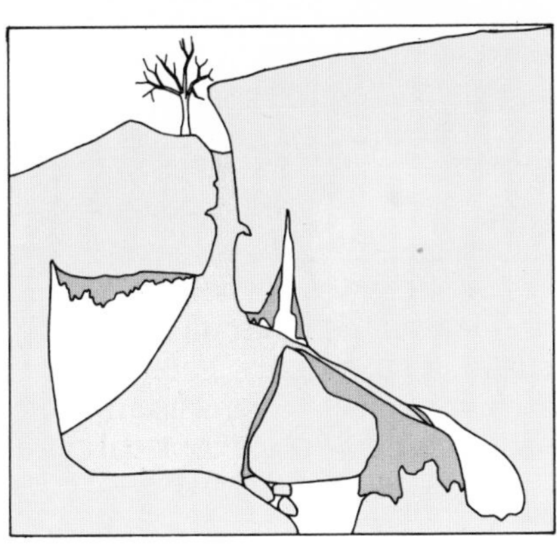

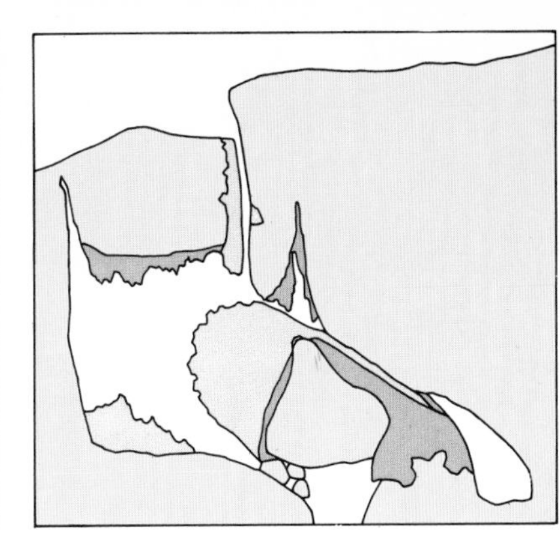

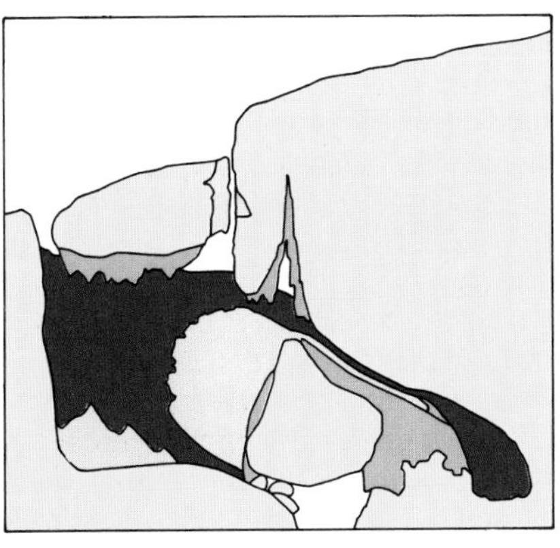

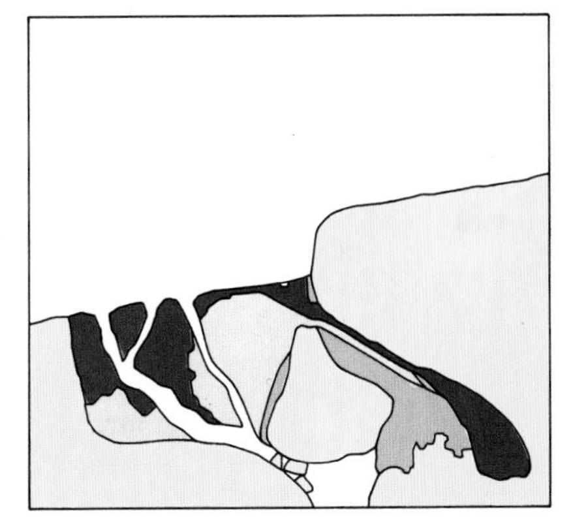

This series of diagrams represents Bob Brain's current theory about the formation of the Swartkrans cave deposits.

1. The initial stage was the hollowing out of the cave by rainwater which seeped through the dolomite limestone.

2. A block became detached from the cave roof and fell to the cave floor.

3. Deposits of travertine accumulated on the cave roof, over the floor block and on the floor of the Inner Cave.

4. A vertical sinkhole opened up above the Outer Cave through which debris, including soil, rocks and bones, fell into the cave. With time, the debris turned into rock: the member 1 breccia.

5. The sinkhole became choked with debris, and the connection between the Inner and Outer Caves was also blocked off.

6. A channel again opened up in the sinkhole and rainwater ran through, eroding away large parts of the member 1 breccia.

7. A new sinkhole developed above the Outer Cave, and the floor block subsided slightly, opening up the way into the Inner Cave. Debris falling through the new sinkhole filled up most of the space in both the Inner and Outer Caves. This eventually turned to rock: the member 2 breccia.

8. Hillside erosion has removed the roof of the Outer Cave, so that the breccia is exposed. Rainwater has opened up irregular channels in the breccia, some of which have later become choked with sediment (not shown). The sediments in these 'channel-fills' vary considerably in age, complicating the picture still further.

deposited and had become rock-like, fluctuations in the water table began to erode the deposit, washing some parts away altogether and leaving a honeycomb of solidified deposit in other parts of the cave. When the cave opened a second time, possibly some half-a-million years later, the new material was deposited onto an eroded surface and filled up the spaces in the honeycomb of old, solidified deposit. The upper layer of the Swartkrans cave is therefore a deceptive mixture of old and new material. Bob Brain considers that this intermittent process of infill, erosion and further infill is common to many of the Transvaal caves. He now says that 'practically every single idea I once had about the cave has had to be modified as I learnt more and more about the complexities.' The learning process has been long and trying: 'I don't think I would have embarked on the project initially if I'd known how much work was involved . . . I wouldn't have had the courage', he admits.

Another notion which Bob Brain has had to modify in the past year or so concerns the way in which the bones found their way into the cave deposits. For quite some time Bob Brain considered that leopards were mainly responsible for the cave burials. The reasoning was as follows: the plateau of the high veld had been largely devoid of sizeable trees, except at the entrances of caves, where accumulated deposits provided sustenance for their deep roots. A leopard frequently retreats to a large tree when it has made a kill, to escape the attention of hyaenas. As the prey disintegrates parts of its body drop to the ground and, where this occurred outside caves, the suggestion was that the bones were eventually swept into the cave mouth.

But when the bone fragments were analysed more closely Bob Brain realized that the proportion of primates – baboons, other monkeys and robust australopithecines – was unusually high. Could this be because the primates were sheltering in the caves during the cold spring and autumn nights of the high veld, as baboons do today? In a research style typical of the man, Bob Brain went at dusk into the Uitkomst cave, 12 kilometres (7 miles) from Swartkrans, and waited for the baboons to arrive. A troop of about thirty animals entered the cave as the light was fading, and began to settle for the night on a raised platform in the cavern. Bob Brain then came out of hiding: 'Although pandemonium broke out in the cave, the baboons could not be induced to leave the place in the dark,' he recalls. Cave sleepers, afraid of the dark, must have been relatively easy prey to an agile carnivore, such as a leopard or a sabre-tooth cat. It is quite likely that hunting hyaenas also brought their meals into the cave from time to time, thus adding to the accumulated bones.

Elizabeth Vrba, deputy director of the Transvaal Museum, has also been analysing the bone accumulations from the Transvaal caves, and her conclusions appear to corroborate those of Bob Brain. She has devised a method of telling whether bone accumulations were the work of predators or scavengers. By studying modern bone accumulations she has established that when a predator, such as a leopard, is responsible the bones will all have come from animals of about the same size and that many young animals will be included. A scavenger, on the other hand, accumulates bones from animals which vary greatly in size.

Using this information she has been able to assess the way in which fossil bones accumulated and she suggests that predators were active at Sterkfontein, Kromdraai and Swartkrans. At Sterkfontein, where Raymond Dart's successor, Phillip Tobias, now works with Alun Hughes, there is an interesting transition in one of the later accumulations. In the layer known as 'member four', where all the *Australopithecus africanus* specimens have

The skull of a robust australopithecine and a leopard skull from Swartkrans. The marks on the hominid's cranium match exactly with the leopard's teeth, a graphic illustration of how this australopithecine met its death. Most of the bones in the South African caves are thought to have come from the meals of leopards and other carnivores.

been found, the pattern of bones is typically that produced by predators, and it is probable that the australopithecines are found there because they were items of prey. In the next, more recent layer, 'member five', there are stone tools and a fossil that might be related to our direct ancestors. Here the bone accumulation indicates the activity of a scavenger rather than a primary predator. The presence of the tools speak of the cave as a living site and it seems likely that the stone-tool users were also scavengers who inhabited the caves and brought their scavenged carrion back there.

One of the major problems with the South African cave deposits is that it is very difficult to say with confidence how old they are. Current ideas put Makapansgat as the oldest, being possibly slightly more than three million years old. Sterkfontein is next, at around two-and-a-half million years; Kromdraai, Swartkrans and Taung follow in order of decreasing age, with Taung possibly being as recent as 700,000 or 900,000 years. The unquestioned fossil riches of the South African caves will assume even more importance when new and eagerly awaited dating techniques become available.

If dating is difficult in the caves, then retrieving the bones is often close to being a nightmare. During the hundreds of thousands of years that the

Alun Hughes points out a fossilized bone in the pink breccia of the Sterkfontein cave.

bones have laid in the cave deposits, they have gradually been transformed into a composite rock known as breccia. Often the breccia is pink or brown, and this rests against the grey walls of dolomite limestone. Erosion through the ages has cut down through the land surface, often removing much of the cave roof, leaving the breccia exposed to open view. For the most part the caves no longer look like caves, but like holes in the ground filled with pink and brown rock. Excavating in the caves is rarely a matter of carefully brushing aside the surrounding matrix: it involves using a hammer-drill to cut out blocks 25 centimetres by 10 centimetres (10 inches by 4 inches) and occasionally the excavators even use dynamite to blast the rock out. 'It's a compromise between speed of extracting the rock and risk of damage to the specimens,' Bob Brain explains.

Once in a laboratory the solid mass of rock and bone goes through many cycles of acetic acid treatment and washing, a process that takes over eight weeks. Ironically, the bone is often so fragile that, as it is gradually exposed by the acid, it has to be coated with a hardening substance to prevent it from disintegrating. This process is fairly hazardous for ancient bones, and specimens of particular importance such as hominid remains are delicately chiselled out with a tiny dental drill, a very slow and laborious procedure.

Elizabeth Vrba began her work on the bone accumulations from the caves, faced by what she describes as 'a mountain of breccia'. Out of that mountain she extracted tens of thousands of fossilized bones of many different species of animal. Prominent among these, however, were antelopes and related animals, and under her analytical eye they began to tell an interesting tale about the Transvaal caves.

Collectively known as the bovids, the antelope group ranges from small forest-dwelling duikers to the wildebeest and gazelles of the plains. As the types of bovids living in any particular area are closely determined by the habitats available, they are potentially useful indicators of prevailing climate in times past. If the bones of mainly plains-dwelling bovids are found among cave accumulations, then there is a good chance that the immediate area around the cave was open grassland.

Elizabeth Vrba looked at the bovid bones in Transvaal cave accumulations and found an interesting pattern. The early layer of Sterkfontein, which contains the *Australopithecus africanus* remains, also has a relatively low proportion of plains bovids. By contrast, the later layer at Sterkfontein, and those from Swartkrans and Kromdraai have a large number of antelopes typical of open savanna, such as springbok, hartebeest and wildebeest. The climate seems to have been considerably drier during this later period so that the forest receded and was replaced by open grassland. This is confirmed by the analysis of fossil pollen which shows that most of the vegetation at this time was grass. It is during this later, drier period that *Australopithecus robustus* was living on the veld.

This climatic switch to drier conditions at around two million years ago appears to be a continent-wide and possibly a global phenomenon. It is just such changes in the prevailing environment that can trigger the generation of new species through natural selection: animals have to adapt to new conditions, and there may be environmental opportunities that new species can exploit. Indeed, one can see in the African fossil record of this time a significant increase in the number of plains animals. It may well be that this climatic change had a significant effect on the later evolution of the human family.

The evidence from East Africa

Although the search for human forebears focused on South Africa for three decades after Raymond Dart recognized the Taung child, the emphasis since the 1960s has swung to a number of extraordinarily productive sites in East Africa. The change is more than simply one of geography: it is a switch to a totally different kind of fossil environment. The fossils excavated in South Africa come exclusively from cave deposits. In East Africa, by contrast, most of the bones of our ancestors were preserved either at the edge of ancient lakes or close to streams or river deltas.

The differences between the two types of site are many. For one thing, cave deposits provide a concentrated source of fossils. When Phillip Tobias or Bob Brain work at their cave sites they simply make a forty-minute drive from their laboratories in Johannesburg and Pretoria to the Sterkfontein valley and there, literally at their feet, is an enormous store of fossil bones: Sterkfontein and Swartkrans are the richest sources of, respectively, *Australopithecus africanus* and *Australopithecus robustus* specimens in the world. Fossil-hunting for my team at Koobi Fora, on the other hand, is a matter of repeated prospecting over more than 1,000 square kilometres (400 square miles) of deposits. We excavate only when something interesting is found. The same is true at Hadar in Ethiopia where Don Johanson, Maurice Taieb and their colleagues have been making spectacular discoveries over the past few years. Prospecting, though over more restricted areas, is also the usual practice at Olduvai Gorge and at the Omo delta in Ethiopia where Yves Coppens and Clark Howell have for many years been searching through the deep river sediments.

It is no accident that the principal sites I have mentioned in East Africa are closely associated with the Great Rift Valley. The floor of the East African valley is dotted with numerous lakes which provide a source of water for many animals. In the past, those animals included our ancestors. More importantly from the fossil-hunter's point of view, the rise and fall of the lake levels and the seasonal flooding of the rivers were responsible for burying and preserving bones. Although this process of entombment does not give such a good concentration of fossils as do the predatory accumulations in the South African caves, the bones that are found are much more often intact. When predators have finished with their meals, the bones are often very fragmented. And the skulls that are not smashed initially may later be broken by the collapse of the cave roof or by the gradual increase in pressure as deposits build up through the years. Many of the hominid skulls that have been discovered in the South African caves are severely distorted. One splendid example of an intact *Australopithecus africanus* from Sterkfontein, popularly known as 'Mrs Ples', survived intact only because the cranium rolled underneath an overhanging rock that protected it from being crushed from above.

Because the Rift Valley marks a weakness in the earth's crust it is punctuated by active volcanoes, and these have left a welcome legacy for prehistorians. As I mentioned in Chapter 2, the chemical components of the volcanic ash act as an 'atomic clock', indicating when each eruption occurred. Fossils buried under the layers of ash are therefore relatively easy to date. Without a good timescale against which to set the hominid fossils that are found, it would be impossible to reconstruct the various phases of human evolution.

Whereas the South African caves provide a concentrated source of fossilized bones, at the East African sites repeated prospecting over large areas is necessary to locate the scattered fossils. This photograph, taken in 1966, shows Louis, Mary and Richard Leakey, together with Phillip Tobias (left), fossil-prospecting near Lake Baringo, Kenya.

The discovery of 'Dear Boy'

The story of the search for fossil man in East Africa took a dramatic turn on 17 July 1959 at Olduvai Gorge. My father and mother were at the Gorge on their annual expedition and were involved in an excavation at a site where, the year before, a single molar tooth had been found. This tooth offered the hope that additional material might be recovered.

On 17 July my father, Louis, was not well and so decided to spend the day resting in camp. Mary was anxious not to waste precious time and so she went off to continue the search that they had been conducting in the Gorge since the early 1930s. Over the years, they had collected an impressive quantity of both fossils and stone artifacts but, despite careful searching, no significant hominid fossils had been found. They both returned each year, convinced that sooner or later their luck would change and the maker of the tools would indeed be found.

On that July morning, Mary was working along the side of a small erosion slope, accompanied by her six Dalmatian dogs, when she noticed an interesting fragment of bone. She bent down and carefully brushed away the soil around the fragment. In so doing, she exposed several unmistakably hominid teeth. The teeth turned out to be part of a complete upper jaw and in due course, excavations resulted in the recovery of hundreds of skull fragments. These were painstakingly reconstructed to provide a remarkably complete cranium.

My parents were so pleased by this amazing reward for their long years of searching that they nicknamed the skull 'Dear Boy'. But because of its huge teeth, a characteristic of these very robust australopithecines, the popular press christened it 'Nutcracker Man'. Although clearly related to the South African *Australopithecus robustus* species, the East African find was even more solidly built. For this reason it was given a new scientific label, *Australopithecus boisei*, in honour of Charles Boise, a London businessman who had contributed generously to my parents' work. Since 1959 a great many more specimens of *boisei* have been found at various sites in East Africa.

When 'Dear Boy' lived at Olduvai, the vegetation would have been somewhat similar to today's, though perhaps rather lusher. He probably inhabited thorn-bush savanna, with fruiting trees and bushes close to the many streams and rivers that fed an ancient lake. Although the view from the lake to the west was of relatively flat plains, to the east there were the volcanic cones of Lemagrut, Sadiman and Ngorongoro. Judging by the fossil record, the diversity of animal and bird life was rich indeed compared to the number of animals that live in Africa today. The rain that fell in the hills and on the volcanic slopes drained into the lake, bringing with it silt. Slowly this silt built up, in and around the lake, eventually forming layers of sediment 100 metres (330 feet) thick. In more recent times, a seasonal river has sliced through these ancient lake sediments, carving out the Gorge with the successive layers of rock exposed in its walls.

The discovery of *Australopithecus boisei* at Olduvai Gorge caused quite a stir in the world of prehistory. Mary explains: 'I was very excited about the find, partly because it was the first *Australopithecus* to be properly dated: it is one-and-three-quarter million years old. This was considerably older than anyone had anticipated. Louis was a little disappointed, though, because he was hoping that we had found an early *Homo*; although *boisei* was important and securely dated, it obviously wasn't *Homo*.'

Within two years of the appearance of *Australopithecus boisei*, however, Louis found what he was looking for. Or rather, Jonathan, my elder brother, found it. Called 'Jonny's child' by the family, this second type of hominid

Mary Leakey with the skull which she and Louis called 'Dear Boy', though the popular press nicknamed it 'Nutcracker Man'. It was clearly related to the robust australopithecines of South Africa, but, unlike them, could be reliably dated by means of the layers of volcanic ash to be found in Olduvai Gorge. A cranium of the same hominid type, found by Richard Leakey near Koobi Fora, is shown on p. 16.

from the Gorge was much more delicately boned than *Australopithecus boisei*, but it was different from the typical gracile australopithecines from South Africa (those which had been called *Australopithecus africanus*) too. This new hominid's brain was substantially bigger than that of either of the australopithecines: it was close to 800 cubic centimetres (49 cubic inches) as opposed to between 450 and 550 cubic centimetres (27 and 33 cubic inches). My father decided that at last the tool-maker had been found, and eventually the new hominid was named *Homo habilis*. The name *Homo* indicated that this hominid was on the line leading to modern humans, while *habilis* means 'skilful' and refers to his ability to make tools. *Homo habilis* lived at the lake almost two million years ago.

The missing years: Hadar and Laetoli

In the previous chapter, David Pilbeam described his evidence and ideas about a possible candidate for the first hominid: an ape-like creature named *Ramapithecus*. The lack of suitable sediments of the correct age means that the last we see of this animal in the fossil record is at around eight million years ago. The discoveries from South and East Africa show that around two million years ago there were at least three hominids in the continent: the gracile australopithecines, the robust australopithecines and *Homo habilis*. The puzzle is, what happened in between *Ramapithecus* and these later hominids?

Until recently the only evidence of the hominid odyssey between eight million years ago and a little more than two million years ago came from

Lothagam and Kanapoi on the southwest shore of Lake Turkana where fragments of a lower jaw and a fragment of an arm bone were found that were thought to be five-and-a-half million and four million years old respectively. These hominid specimens are too incomplete to provide any useful conclusions and it is perhaps wise to leave them on one side until they are joined by other fossil material of the same age. Fortunately, two of the most remarkable research sites to be exploited recently – Laetoli in Tanzania and Hadar in Ethiopia – have yielded some magnificent evidence relating to the period between four million and two million years ago.

'Lucy' was a small female australopithecine who lived three million years ago, beside a lake in what is now Ethiopia. With forty per cent of her skeleton recovered, she is the most complete specimen of an early hominid ever found. The shape of the pelvic bone shows that she was female, while the leg bones indicate that she walked upright. Her teeth suggest that she was about twenty years old when she died.

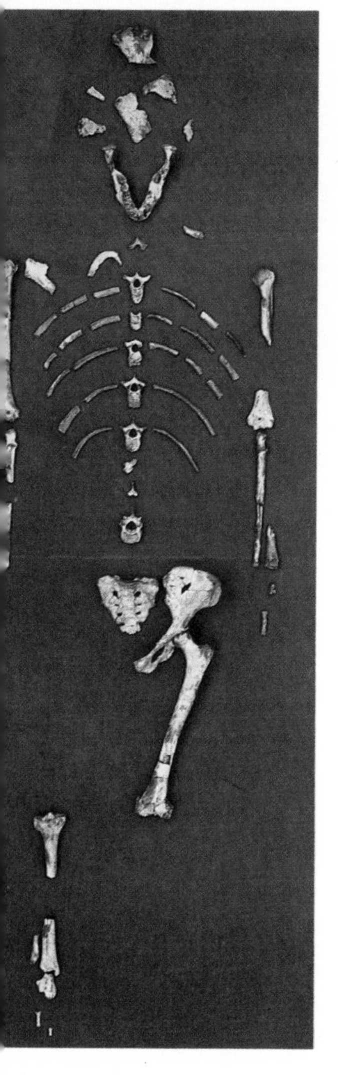

In addition to those fascinating ancient footprints described in the previous chapter, my mother and her colleagues have also discovered twenty-three fragments of hominids which are three-and-a-half million years old. As one goes farther and farther back into our evolutionary past, it inevitably becomes more difficult to distinguish one hominid species from another. The closer one is to the 'rootstock', the more the separate species look like each other. Mary's Laetoli fossils present just such a problem: are they *Australopithecus*, are they *Homo*, or are they something quite different? 'They have many characteristics that are very similar to those of *Australopithecus*,' she says, 'but I consider that they are the only possible candidate for an ancestral form of *Homo* at this particular date.' So Mary believes them to be a primitive species of *Homo*, possibly ancestral to *Homo habilis*. Not all prehistorians would agree with this position, but I feel that a good case can be made out for its support.

Since 1972 Don Johanson, Maurice Taieb, Yves Coppens and their colleagues have been prospecting under difficult conditions at Hadar in the Afar Triangle, Ethiopia. As at Koobi Fora, the Hadar sediments were laid down millions of years ago along the margins of a fluctuating lake. Unlike Lake Turkana, the Afar lake has long since disappeared, but fortunately for the researchers, the Awash River provides a source of water for cooking and drinking, and a place to bathe. Hadar, which means 'dry river', has proved to be one of the most abundant sources of fossils in Africa for the period between two-and-a-half and three-and-a-half million years ago.

In November 1975, Don Johanson and his colleagues made a find that, even by Hadar standards, was stunning: they came across a concentration of several hundred bones that represented at least thirteen and possibly more individuals. The discovery posed many problems, not least of which was how the bones of so many individuals came to be fossilized in the same place at the same time. 'It looks to me to have been some ancient catastrophe,' Don Johanson says. 'There's no sign that the bones lay around for a while, forming some kind of accumulation over a period of time. They must have been moved around quite a bit, possibly in a stream, because many of the bones are broken. But it looks to me as if all these individuals died either at the same time or within a short interval of each other.' Of the thirteen individuals in the group, four are infants below the age of five (the teeth give this information) and the rest are probably a mixture of adults and juveniles.

The discovery of the 'family group' as it has often been called, came only a year after one of the palaeontological surprises of the century. While Don Johanson and a colleague, Tom Gray, were trying to relocate an interesting geological site they had noticed the previous season, they found the first fragments of what eventually turned out to be forty per cent of a single, small hominid skeleton. The skeleton became famous under the name of 'Lucy', inspired by the Beatles' song 'Lucy in the Sky with Diamonds' which someone was playing on a tape one night in camp shortly after the skeleton turned up.

Left: The camp beside the Awash River where the Hadar team live. Since they began work here in 1972 they have retrieved the remains of between thirty-five and sixty-five early hominids.

Right: Only a year after the finding of Lucy, came the discovery of the 'family group': a collection of bones which represent at least thirteen and possibly more individuals. As the size of the new find becomes apparent, Don Johanson holds up yet another bone for his colleagues working nearby to see.

The Hadar team, which unfortunately had to abandon research for a few years because of the war in Eritrea, has now retrieved at least thirty-five individuals, and possibly as many as sixty-five, from those lake deposits. The question is, what species are they? Initially, Don Johanson, Maurice Taieb and Yves Coppens published a paper suggesting that some of the fossil hominids were probably *Homo*, while others were *Australopithecus*. 'But now, after studying the fossils very carefully for a long time, I've changed my mind,' says Don Johanson. 'Instead of there being two separate species at Hadar, I believe there is just one.' With Yves Coppens and Tim White, a colleague from the University of California, Don published a paper suggesting that the Hadar hominids belong to a species which they named *Australopithecus afarensis*, after the Afar Triangle. They also said that the Laetoli hominids belonged to this species. But their most significant claim is that *Australopithecus afarensis* is the rootstock ancestral to all subsequent hominids, that is, to the australopithecines as well as the *Homo* species.

There is no doubt that the Hadar hominids are absolutely fascinating. Owen Lovejoy has been studying the bones in detail and he comments: 'They look incredibly primitive above the neck and incredibly modern below. The knee looks very much like a modern human joint; the pelvis is fully adapted for upright walking; and the foot, although a curious mixture of ancient and modern, is adequately structured for bipedalism. Some of the bones in the feet are slightly curved, and look rather like the bones you'd expect to see in its ancestor who climbed trees. But I believe that the curvature in the *afarensis* foot bones is well suited for walking on soft, sandy terrain: it probably inherited the curved feet from its tree-climbing ancestors, but the shape has been made use of in a different way.'

Don Johanson is also struck by the combination of ancient and modern: 'These things have clearly made the initial important evolutionary move – they walked upright. But the teeth and the jaws are very similar to those of *Ramapithecus* in Pakistan. And they had very small brains, less than 400 cubic centimetres (24 cubic inches). It's amazing to see something so very primitive at such a relatively recent age. We believe that *afarensis* is ancestral to the later hominids because it is basically primitive but has some modern features emerging.'

One facet of human evolution that the discoveries at Laetoli and Hadar highlight is the early development of upright walking. As Don Johanson explains, 'It's clear from these fossils that upright walking happened long before brain expansion. Hominid brains don't show any striking signs of getting particularly big until two to two-and-a-half million years ago, and yet these creatures were bipedal at least a million years before that.'

The claims made by Don Johanson and Tim White regarding the finds from Hadar and Laetoli have provoked a good deal of discussion. Some people agree that the fossils from the two widely separated locations do represent one species, and that this species, *Australopithecus afarensis*, was ancestral to the hominids that lived around two million years ago. Others, myself included, interpret the evidence differently. The arguments are rather esoteric and centre on subtle bumps and curves in the fossilized bones, but, simply stated, I believe that the Hadar collection is composed of two species. One is bigger than the other, and there are also some subtle, but important, anatomical differences between them.

According to Don Johanson and Tim White, the bigger individuals in the whole group are male *afarensis*, and the smaller ones are females of the same species. Owen Lovejoy says that the *afarensis* 'males' are around sixty to seventy per cent bulkier than the *afarensis* 'females'. This is slightly less than the difference between male and female gorillas, but as these fossil hominids are so small – their weight would probably have been between 25 and 55 kilograms (55 and 120 pounds) – the suggested disparity between males and females is fairly dramatic. I believe, therefore, that it is unlikely that the Hadar hominids are all of one species.

I envisage two species of hominid living at Hadar just over three million years ago: a larger species which were a primitive form of *Homo*, and a group of smaller hominids belonging to a previously unknown *Australopithecus* species. The *Homo* line had to arise at some point in time of course, but I suspect that the instant is farther back in time than the Hadar and Laetoli deposits. Given the well developed nature of *Homo habilis* at around two million years ago, and what I see as the diversity of hominids at three-and-a-half to three million years ago, it seems a fair guess that the *Homo* line may have evolved initially as long ago as five million years. There could have been a common hominid stock at that time, possibly *Ramapithecus* or a derivative of it, from which the emergent *Homo* and *Australopithecus* species could have sprung. Inevitably, the question is not going to be settled until there is a great deal more tangible evidence from this period.

Meanwhile it is probably wise to keep something of an open mind about the relationship of the various hominids to each other, especially during this early period. There has long been a tradition in human prehistory of viewing evolutionary progression as something approaching a straight line. Indeed, there was once a popular view, still held by some, that all hominids were members of the same lineage, each on its allotted rung of a ladder that ultimately reached the level of *Homo sapiens*. The processes of evolution are, in fact, much more complex than this notion allows for.

Hominid evolution was certainly not a single steady movement along a path of gradual 'improvement'. Nor, probably, was the story of human prehistory peopled by the few hominid species scientists recognize today in the fossil record. The passage of time from eight million years onwards almost certainly saw a number of geographically separated species, most of which thrived briefly but in the long term left no descendants. Some, however, were more successful, and these are the lines of hominid descent which prehistorians are trying to trace. Stephen Jay Gould compares the old simplistic view of human evolution with the modern, biologically realistic view in the following words: '*Homo sapiens* is not the foreordained product of a ladder that was reaching toward our exalted estate from the start. We are merely the surviving branch of a once luxuriant bush.'

The coexistence of the early hominids

Leaving aside the problem of where precisely the hominids of two million years ago came from, the next important question to ask is: what were they doing? How did the australopithecines and their *Homo* cousin make a living and what was their ecological relationship with each other?

Much of what can be said about the behaviour of ancient hominids is, perforce, guesswork. One can, however, begin with some firm facts. For example, we can now say that the australopithecines definitely walked upright, whereas a few years ago this was still a matter for speculation. Owen Lovejoy, who has made a special study of the way hominids get around, concludes that, although the details of the anatomy of the australopithecines show shades of variation from modern humans, the practical implication is that they walked around on two feet with as much ease as we do today.

The adoption of upright walking, combined with an analysis of the environments in which hominid bones are found, implies that our ancestors of two million years ago lived in much more open country than did their forebears. Not that they would have entirely left their arboreal habits behind them. They were probably capable of climbing trees in times of danger, and it is possible that they slept in trees, as baboons do, in areas where there were no caves for shelter.

Primates, for the most part, are highly social creatures, living in troops and having intense personal interactions. Modern humans are no exception to this, and neither were the hominids of two million years ago. The exact form of their social life would probably have been determined to a large extent by exactly how they made a living, what foods they ate, and how these food resources were distributed in the environment. It is, however, very difficult to say with certainty anything about the food eaten by our ancestors.

The teeth which represent so many hominid discoveries can certainly reveal something about diet. Consider the robust australopithecine with his massively buttressed jaws, huge millstone-like molar teeth and tiny front teeth: what do these characteristics imply? In spite of his nickname, 'Nutcracker Man', biomechanical calculations show that the pressure *Australopithecus boisei* could apply between his teeth was exactly the same as that which modern humans are capable of with teeth one quarter the size. The same is, of course, true for *Australopithecus robustus*. The reason for the huge dental apparatus was not to break down particularly hard or tough foods but, perhaps, to process a large quantity of food, which implies that the food was of fairly low quality.

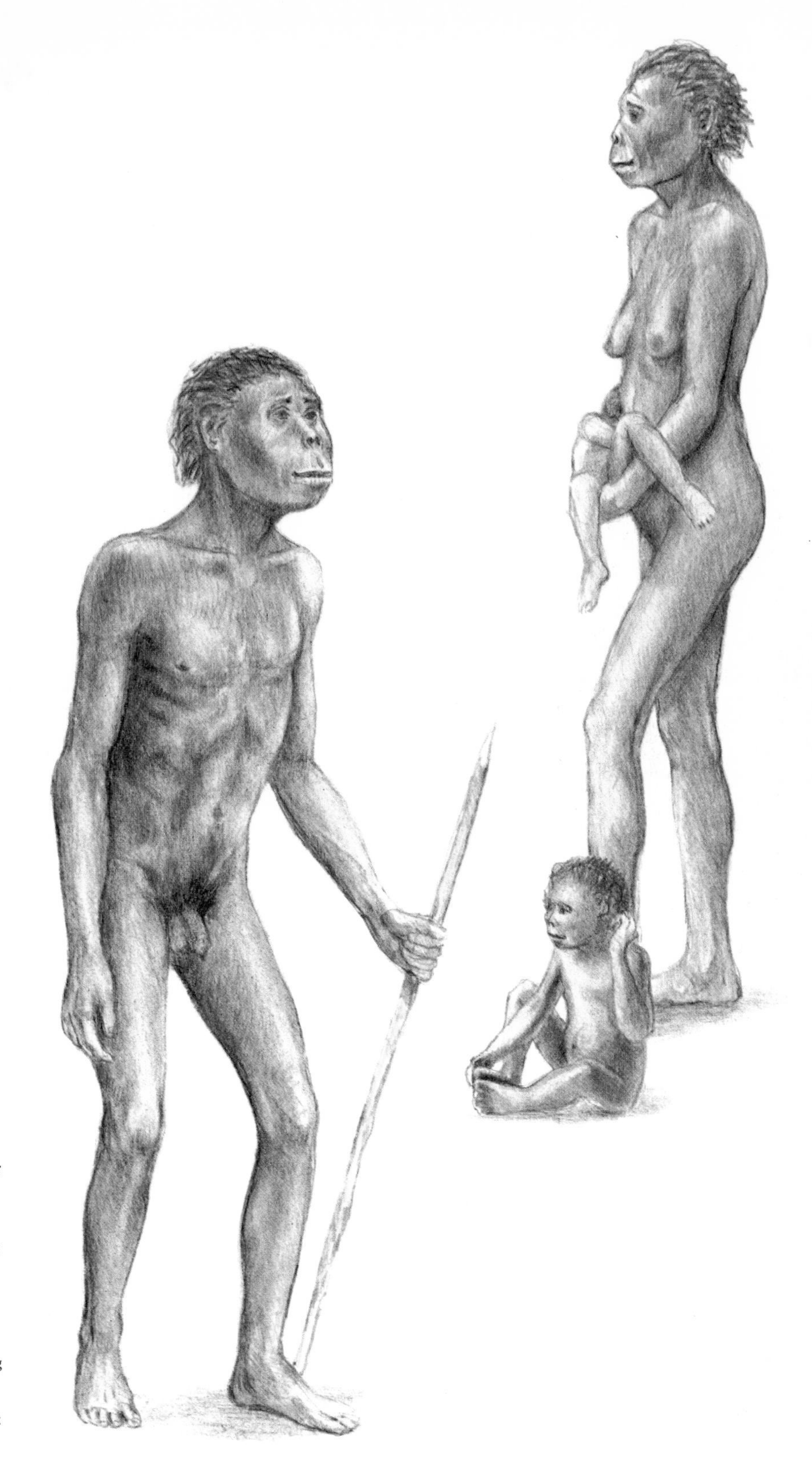

The robust australopithecines (far left), the gracile australopithecines (left) and *Homo habilis* (right) were all inhabiting parts of Africa two million years ago. It is almost impossible to show that they were living in exactly the same places at the same time, but this may have been the case. If so, how did they coexist without coming into direct competition with each other? One suggestion is that they utilized different food sources, with *Homo* including more meat in its diet. The australopithecines eventually died out, leaving only *Homo*, our direct ancestor.

Some years ago John Robinson, Robert Broom's principal colleague, proposed a dietary scheme for hominids. The robust species, he suggested, was primarily a vegetarian animal, living on grasses, roots, shoots and berries. The gracile australopithecine, *Australopithecus africanus*, was thought to be more of a generalist, adding lizards, eggs and even small mammals to its diet, while *Homo* had shifted even farther away from the vegetarianism typical of primates and had developed a diet in which meat played a considerable part.

This dietary scheme is worth repeating because it highlights the need for some form of ecological separation between the hominids. Similar species that make a living in a similar way cannot live in the same environment without facing unacceptable competition from each other. As far as one can determine from the fossil evidence, these hominid species were indeed very similar. Some kind of separation is therefore necessary and John Robinson's dietary spectrum or something like it is very plausible. But the separation need not necessarily be a matter of diets. If, say, *Australopithecus robustus* spent most of the time in less wooded areas while its gracile relation lived largely on the forest edges, that too would effect an ecological separation, although they would be inhabiting the same general area.

One can infer from the evidence of the South African caves regarding changes in climate that the ecological separation of the australopithecines could have been even more complete. Suppose the gracile australopithecines occupied areas where the climate was moist enough to support a fairly substantial covering of bush and trees, whereas the robust australopithecines favoured a drier terrain. A single geographical area could switch from one type of vegetation to the other over a few hundreds of years given only minor climatic fluctuations. The inference drawn from the fossil record that the two hominids lived in the same location at the same time would then be incorrect. Such subtle changes will inevitably be difficult to document.

Fortunately, the shape and size of the teeth are not the only clues they can offer about past diets. Alan Walker, now working at Johns Hopkins University in Baltimore, is using an electron microscope to examine the minute scratches that are left on tooth enamel and dentine by food as an animal chews or crushes it. He is building up a 'library' of typical patterns for herbivores, fruit-eaters, bone-crunching carnivores, non-bone-crunching carnivores and ground-living omnivores. The hope is that when he analyses a sample of fossil hominid teeth he will be able to place the hominid into one of these categories. Although that is the objective, it is likely that the answer will be somewhat uncertain, since most animals tend to eat many different things and defy classification into such simplified categories.

This promising work is in its infancy as yet, but there are some preliminary results that are very surprising. For example, Alan has analysed a number of *Australopithecus robustus* teeth and they fall into the fruit-eating category. More precisely, their teeth patterns look like those of chimpanzees. Now, although chimpanzees depend heavily on fruit (not the brightly coloured, succulent items we buy in shops, but hard, thin-skinned wild fruits) they also eat a host of other things, including ants and termites, sap, birds' eggs and small animals. So, to say that *Australopithecus robustus* has a tooth-wear pattern similar to that of a chimpanzee does not define his diet very closely. It does, however, exclude grass-eating, bone-crunching and root-eating, all of which would have left distinctive scratches and pits on the enamel.

This result for *Australopithecus robustus* is something of a surprise, but even more surprisingly the few examples of *Ramapithecus* teeth that have been put under the microscope show exactly the same pattern. And the teeth of *Homo habilis* also have the smooth enamel typical of a chimpanzee. Alan says: 'I'm inclined to believe these results, simply because they are *not* what I would have predicted.' Then, when he looked at some *Homo erectus* teeth, he found that the pattern changed: 'They were nothing like the other hominids I'd seen', he says. 'They looked just like bushpigs.' Bushpigs are omnivores eating more or less anything they can find: roots, bulbs, grass, bark, berries, fruit, seeds, fungi, eggs, insects, reptiles, carrion and occasionally young birds. They 'root' for their food, digging into the earth with their snouts. This means that they pick up a great deal of grit with their food, which leaves huge scars in the enamel. 'It looks like a layer of concrete that someone's attacked with a sledgehammer,' Alan says.

Homo erectus may have picked up grit on roots he unearthed: certainly, the discovery of roots and tubers as a source of food would be a major economic breakthrough because in arid areas there is often five to ten times more food below ground than there is above. The grit may also have been adhering to meat that had been lying on the ground, but there are no signs of bone-crunching on the teeth, so it is difficult to make comments about meat-eating from this evidence.

Alan Walker's tooth work is extremely promising, although the stories etched into the tooth enamel may simply be too indistinct to differentiate clearly the lifestyle of the three hominids. But if one considers the brains of these ancient hominids one can see what might have been the differentiating factor between *Homo* and the australopithecines: intellect. The fossil skulls of the australopithecines suggest that there was little or no change in the size or shape of their brains over a period of more than one million years. On the other hand, one of the oustanding characteristics of the early *Homo* species was the increase in the size of the brain. With a sharper wit *Homo* may well have devised ways of finding new foods, thus broadening their economic base, and they may have developed a social system to support this new type of economy: these topics will be considered in the next chapter.

5 The Litter of the Past

People have always dropped litter. We see evidence of this all around us, in our city streets and rubbish dumps. Our museums display objects salvaged from the litter of past eras: Roman coins, Egyptian pottery, Ming china, Inca textiles. The list is endless, and each item tells us the same general story: that humans make things, use them, and then dispose of them, either casually as rubbish or occasionally as part of a ceremony such as the burial of an important person.

Litter of the past is the basis of archaeology. The coins, the pottery, the textiles and the buildings of bygone eras offer us clues as to how our predecessors behaved, how they ran their economy, what they believed in and what was important to them. What archaeologists retrieve from their excavations are images of past lives, but these images are not pulled ready-made from the ground: they are pieced together slowly and painstakingly from the information contained in the objects found. Archaeology is a detective story in which all the principal characters are absent and only a few broken fragments of their possessions remain. Nevertheless it has been possible, in many cases, to fill out the details of the story. We know, for instance, how the Incas operated their highly structured welfare/feudal economy and how the Romans organized their sprawling empire.

Although the detective work involved in reconstructing such civilizations can be difficult, those of us who are concerned with the very early stages of human evolution look with envy at the abundant evidence about these recent periods. One of the outstanding features of human history is the steady increase in the production of objects, such as tools, clothing and artificial shelters. As we search back through time, towards our origins, we find an ever-thinning archaeological record. Somewhere between two and three million years ago, human artifacts disappear from the fossil record entirely. The task of discovering what our ancestors did in their daily lives therefore becomes more and more difficult the farther into the past we search.

Investigating our ancestors' behaviour

If prehistorians were to concentrate only on fossil bones, they might deduce something of what these creatures looked like and how their evolutionary paths were linked, but they would discover little about their behaviour. Alan Walker once said at a meeting on human origins: 'If all we were to learn from prehistory was a list of species in time and space then we should give up. That would be trivial. We want to know how our ancestors lived; how they survived together; what their sex life was like; we want to know how they *behaved*.' In other words, we want to know what, eventually, made them human.

Glynn Isaac, a close friend and colleague at the University of California,

Excavating a site near Lake Turkana, known as site 50, which was probably the camp-site of a group of early hominids. The litter they left behind offers important clues about their way of life.

has been investigating ancient behaviour at Lake Turkana for the past ten years. Glynn describes an important shift in the approach to work on early sites: 'In the old days the "prize" sites were those with masses of material – lots of stone tools and bones – that would provide good collections for museums. Such sites attracted archaeologists like iron filings to a magnet. But now we are interested in tools, not as objects in themselves, but as part of a social context. We want to be able to use what we find on archaeological sites to test hypotheses we have about the actions of protohumans who lived one-and-a-half million years ago.'

Wherever researchers have looked for fossils of human ancestors they have also searched for tools. So far it is clear that fossil hominids appear in the record *much* earlier than their tools: earlier by at least one million years, and possibly by as much as two million years. Among the oldest African remains that are unquestionably hominid are fossils close to three-and-three-quarter million years old from Laetoli. And, as I suggested in the previous chapter, the first creatures on the line of descent leading to *Homo sapiens* may have evolved a good deal earlier than that.

In contrast, the earliest stone tools known at present appear in the archaeological record two million years ago: these small quartz flakes come from excavations in the Omo Valley. However, there is a rival claim for the oldest tools, from Hadar in Ethiopia, where French archaeologist Hélène Roche discovered crude stone flakes that were undoubtedly made by hominids. Their age is a little uncertain, but they could be as old as two-and-a-half million years. At later stages in the archaeological record, there are far more sites at which stone tools have been found and the quantity of artifacts at each site also increases. More interesting, and more significant in terms of the story of human evolution, is the fact that the tools are often intermingled with the remains of animals.

There is a small erosion gully about 25 kilometres (16 miles) inland from the Koobi Fora camp where, in 1969, I came across fragments of fossilized hippopotamus bones which were apparently being eroded from a layer of volcanic ash. Jack Harris, John Onyango-Abuje and Glynn Isaac supervised the subsequent excavation of the site, which eventually uncovered over a hundred stone tools. The tools were a mixture of sharp flakes, larger stones from which flakes had been struck, and a rounded river pebble that looked as if it had been used as a hammer stone with which to strike flakes from the cores. These remains are almost two million years old.

One interpretation of this site is that the tool-makers killed and then butchered a hippo. It is, however, more likely that a band of hominids stumbled across the recently dead animal and, taking advantage of their lucky find, cut off pieces of meat. But, as Glynn always insists, one cannot completely dismiss the possibility that the bones and stones had nothing whatever to do with each other. It could be that a group of hominids chose to sit in the comfortable sand of a dry river delta where they knapped stones and then departed leaving their litter behind them. Later an old hippo came along and died, by chance, among the hominids' stone debris. There is no way to distinguish between this last hypothetical sequence of events and the first two interpretations.

About a kilometre (half-a-mile) north of where the hippo died, a second interesting collection of bones and stones has been excavated from a layer which covered them almost two million years ago. The stone tools were lying on what, at the time they were made, was the sandy bed of a seasonal stream channel. This channel would undoubtedly have been bordered by trees and bushes, offering shade from the glare of the tropical sun and, perhaps, refuge in case of danger from predators. Fresh water could have

A seasonal stream near Koobi Fora today. A similar sandy stream channel would have offered shade, a comfortable resting place and a source of fresh water to our ancestors.

been reached by scooping through the sandy riverbed, giving another possible reason for hominids to have chosen this spot as a camp-site.

Archaeological finds strongly suggest that the place was indeed used in this way. Among the stone tools there lies a scattering of bone fragments. They are mostly small and corroded but they are identifiable and belong to a varied collection of animals, including hippopotamus, giraffe, pig, porcupine, waterbuck, gazelle and wildebeest. We cannot be absolutely certain that these bones are refuse from meat which the ancient tool-makers carried here and ate, but this is the most likely explanation. The scatter of bones and stones would suggest that a hominid band stayed here for a while, eating meat which was probably scavenged from dead animals. After a few days they probably left to search for food elsewhere.

Impressions are clearly important when one is attempting to fit together

the pieces of this ancient puzzle, but even educated impressions are of limited use. There can be no substitute for rigorous analysis, and this is what Glynn Isaac and his colleagues are doing at yet another site near Koobi Fora.

Site 50

A hot dusty two-hour drive northeast from Koobi Fora brings you to the Karari escarpment, a low ridge running northwest to southeast. A river, dry for much of the year, meanders by on its 20 kilometre (12 mile) journey to Lake Turkana. On the river bank is a camp of green canvas tents that is home for archaeologists, geologists and palaeoanthropologists for periods from just a few weeks to several months. The stands of *Acacia tortilis* and other trees make the location strikingly beautiful and cast shade over the camp.

A short walk takes you from the relative cool of the riverside vegetation into gently rolling arid countryside dotted by flat-topped acacias and *Comifera* with their viciously thorned silvery branches. Plants small and large cling to a precarious existence in what is typical East African thorn-bush savanna. It is in this arid landscape that site 50 is found, a substantial excavation on the east side of a hillock. One-and-a-half million years ago the scene here was different. A smooth, relatively lush flood plain ran for 15 to 20 kilometres (9 to 12 miles) to the hills in the east, and the same distance to the lake in the west. One of the many rivers lacing the flood plain ran close to site 50. Although it appears to have been a substantial channel, like today's river at Karari it was probably empty for parts of the year. But when a small group of hominids visited the spot one-and-a-half million years ago there may well have been water running in the river or at least standing in pools, for a piece of catfish skull has been found among the remains uncovered at the site.

The individuals who occupied this place on the inner bank of a bend in the river, left behind them more than 1,500 stone fragments. These fragments occur in two main concentrations. About 2,000 fragments of fossilized bone have also been recovered from the site, again concentrated in two areas. Some time after the site was deserted, probably not very long after, the river gently flooded its banks, covering the bones and stones in fine silt. A distant volcano erupted, throwing whitish-grey ash into the atmosphere, and some of the settled ash was layered on the bank of the river over the abandoned camp-site. As year after year passed, and millennium after millennium, more and more silt was deposited. Then the forces of deposition were replaced by forces of erosion. Eventually, one part of the surface on which the camp-site had existed was again exposed as a small erosion gully cut into it, and thus the site was discovered.

Paul Abell first spotted the site some years ago, but it was left unexcavated until March 1977, when Jack Harris decided to determine whether the trail of stone flakes and fossilized bone fragments coming from the erosion gully signalled a worthwhile site. A test trench revealed more stones and bones from what appeared to have been an ancient living floor. The uncovering of site 50 and its implications began there.

There are very few living sites yet excavated from this distant point in human prehistory, for their initial entombment must have been a precarious process, and the chances of their later exposure and discovery are very slight. One of the most outstanding sites of this type in East Africa is at Olduvai Gorge. The 'Zinj' site, as it is known, yielded thousands of bone fragments and stone tools. There is a wealth of material at the site, but, ironically, there is too much for some purposes. Glynn explains: 'We thought the best chance of beginning to understand the past clearly, in

The landscape to the east of Lake Turkana is one of stark hills and gullies, punctuated by the delicate pink of the desert rose. As streams cut erosion gullies through the sedimentary rocks, ancient camp-sites such as site 50 have been revealed.

The initial task at site 50 was to carefully remove the deposits that covered the camp-site. This work was carried out by an expert group of Kenyan excavators.

terms of our ancestors' economic behaviour, was to analyse a relatively simple living site. The "Zinj" site has masses of material; hominids occupied the spot, probably seasonally, many many times; it's not so easy to sort out. Site 50, on the other hand, looked more promising. In fact, it turned out to be much better than we'd hoped: it was probably occupied for a fairly short period and it was buried rapidly but gently so that it was only minimally disturbed.'

So, in 1977, the work began. Glynn's team of researchers set out to determine exactly what the environment was like one-and-a-half million years ago and to decipher the hidden messages in the bones and stones. The first process, of course, was to carefully remove the deposits that covered the ancient land surface, a job carried out principally by an expert group of Kenyan excavators. This exposed the ancient river bank over which was strewn the litter that our ancestors left behind. But how can one be certain that this 'unpromising jumble of bones and stones', to use Glynn's phrase, is really the result of hominid activity? How does one know that the stones did not just crack in the heat of the sun or shatter under the hooves of passing animals? And surely the bones could be the leftovers of a hyaena's meal? These are very valid questions and ones that Glynn and his colleagues are now tackling.

A closer look at the bones and stones

Technology is to a large extent determined by the raw material available. Here at the Karari escarpment, the solidified lava of nearby volcanoes was the most common rock used to manufacture simple stone implements. As chunks of lava are rolled along riverbeds during floods they are gradually smoothed, yielding pebbles of convenient size from which to strike flakes. Much less commonly used were chert and ignimbrite, which are also the products of volcanic eruptions. Although these are more desirable than lava in that they give sharper flakes, they are usually available only as small pebbles.

Two of Glynn's students, Ellen Kroll and Nick Toth, are principally concerned with interpreting the scatter of stone fragments at site 50. To help them in the task of interpreting the fragments, Nick Toth has taught himself the skills of stone knapping. 'I've made around 1,000 pebble tools since I started research on this general problem about three years ago,' he says. 'Each pebble "tool" probably has about twenty-five flakes and fragments taken from it, leaving a long jagged edge on one side while the opposite side remains smooth. By now I have a strong feeling about the breaking characteristics of these pebbles and I can judge what kind of flakes have come from a core. I can therefore determine pretty accurately whether a flake has been artificially struck or is instead the result of a natural fracture.' Nick Toth is confirming his intuitive judgment with a systematic computer analysis of the flakes. The combination is producing some new insights into the intentions of our ancestors when they made tools some one-and-a-half million years ago.

For many years the traditional work on stone tools of this era has focused on the classification of implements into various classes: end-choppers, side-choppers, discoids, protobifaces and so on. Generally the classification depends on the shape of the tool, but sometimes a function is implied by the classification, such as 'scraper'. In the manufacture of all these tools a scatter of small flakes is produced and these are often described as 'debitage' or 'waste'. Nick Toth and others now question this view. Experience of the past few years, both in making tools and in using them, has produced much more respect for the humble flake. Indeed, Nick goes so far as to suggest that the flake was the primary purpose of most stone knapping in the earliest stages of tool-making.

In the hope of gaining more insight into the stone fragments found at site 50, Nick Toth taught himself to make stone tools. In the three years since he began stone knapping he has made about 1,000 tools and has, on occasion, used them to cut up meat. His experiences are immensely valuable in interpreting the finds from the excavations.

'When I take a cobble and try to obtain as many flakes as possible, and then discard the core when it's too small or too uncomfortable to hold, I finish up with the entire range of so-called pebble tools that have been found in the Koobi Fora area at this time range.' Nick knows what tool shapes have been uncovered at the various Koobi Fora sites, and admits there must be a possibility that he leans, however unconsciously, towards these forms while he is striking flakes. This would, of course, invalidate his claim for the flake. But he has some support for his suggestion: 'When I asked Loriano, our Dassanetch assistant, to make a large number of flakes [this was for another project], he too finished up with "end-choppers", "side-choppers", "discoids" and so on, as a by-product of his major task.' This does not prove anything about past technology, but, as Nick says, 'it's very suggestive.'

If, for the moment, we accept that early humans of this time were including meat in their diet, then the application of stone-tool technology could have been very significant. 'Chimpanzees and baboons eat fresh meat as often as they can get hold of it,' Nick explains. 'But the prey are always small animals: for baboons it may be spring hares or infant Thomson's gazelle; chimpanzees, being bigger creatures, are a little more ambitious, including baby baboons and colobus monkeys on their menu. Neither chimps nor baboons use tools when they are catching their victims or when they are dismembering them: it's all done with their hands and teeth. I'm convinced that the small stone flake was a major technological breakthrough for our ancestors around two million or more years ago.'

If you examine a stone flake of the sort to which Nick refers, it is easy to be unimpressed: it is 2 or 3 centimetres (about an inch) long with a single sharp edge and nothing much more to mention. 'But,' insists Nick, 'you can do almost every aspect of butchering with stone flakes: you can dismember an animal as large as a cow with them. You need a large number of them, you have to be patient and your hands get sore, but it's possible. I know, because I've done it.' Simple lava flakes can even penetrate the hide of an elephant, which is almost 2·5 centimetres (1 inch) thick. These small stone implements could therefore immediately have given our ancestors access to the meat of large animals, an extremely valuable food resource which was not available to them previously. It could have signalled the beginnings of a new economic order that was to separate our ancestors from their non-human primate cousins. Nick Toth's use of the term 'breakthrough' is by no means an exaggeration.

Later tools, dating from around one-and-a-half million years onwards, include much larger items: handaxes and cleavers. This technology is generally called the Acheulean. These handaxes and cleavers can, in basic terms, be viewed as extra-large flakes: they have more form to them, particularly the teardrop-shaped handaxes, but their principal benefit is that they have a much enlarged cutting edge. 'I can work much faster with a cleaver or a handaxe than I can with a small flake,' Nick Toth says. 'I can make longer sweeps with them, I can cut deeper into the meat, and my fingers don't get as tired.' The breakthrough required for producing cleavers and handaxes is simply the skill and foresight to strike a big flake from a big core. A cleaver is roughly wedge-shaped with an 8- or 10-centimetre (3- or 4-inch) long cutting edge, and requires only a little finishing. A handaxe, by contrast, can be a work of art, and involves careful finishing to give two sharp edges.

The 1,500 or so stone fragments on site 50 represent a simple flake technology; there are no handaxes or cleavers. Nick Toth has examined the material and he has identified about sixty cores. 'This represents about an hour's work for an experienced stone knapper,' he suggests. Close study of the cores reveal that they are all the product of the same idiosyncratic process: whoever struck the flakes from the cores did so while holding the target cobble in the hand. Furthermore, they struck flakes from one side of the core only. 'Perhaps it was just one person who made all the tools on this site,' Nick proposes, 'or maybe there were several individuals all following the same cultural pattern.'

When someone sits or squats while making stone implements, fragments shoot off in many directions. There is, however, a particular pattern produced by such knapping activity, the most obvious being that the flakes and splinters land in front of the knapper's body, not behind. It is for patterns such as this that Ellen Kroll and other team members have been searching in the apparently random scatter at site 50. This has involved, among other things, attempting to find stone fragments that were struck from the same cobble, or 'conjoining' the fragments.

Ellen explains the challenge of trying to put the stone pieces back together: 'It's like a jigsaw puzzle, but there's no picture, the edges are all kinds of shapes, and lots of pieces are missing.' As if that were not difficult

Baboons eat fresh meat whenever they can get hold of it, as do chimpanzees, but early hominids lacked their sharp canine teeth and would have had difficulty in dismembering even small animals. Simple stone flake tools, however, can cut through the hide of an elephant, and could have given them access to the meat of large animals as well as small ones. Their invention may have opened up a new way of life for our ancestors.

enough, occasionally two pieces that do fit together may be entirely different colours, simply because they have rested in neighbouring soils with slightly different chemical properties. In spite of these daunting problems, Ellen and other members of the team have succeeded in fitting together more than 150 pieces of the jigsaw: sometimes it is just two fragments that fit, sometimes more. From one of the concentrations at the site all the flakes that came from a 'unifacial chopper' have been found; so far the core is missing, but there are still hopes that it might be buried under a part of the site not yet excavated. The patient jigsaw work has also yielded a core with two flakes and a second set of flakes lacking its parent core. Given the enormous problems of this type of work, not least of which is the fact that some pieces of the jigsaw are certain to be scattered and some of them lost, both while the site is occupied and before it becomes covered by a layer of deposits, these results can be judged as very good.

Luckily for the archaeologists, the site appears to have been relatively undisturbed during the process of burial. The degree of disturbance can be estimated to a certain extent, and this work is being undertaken by Kathy Schick, another member of the Karari team. Kathy takes collections of stone tools – usually about 200 items in each, consisting of cores, flakes and bigger cobbles – and sets them out in various places, very much as they might have been left by a prehistoric band of hominids. She then leaves them, and waits to see how the different elements in the tool assemblages are affected by the forces prevailing in the various locations she has chosen.

For instance, she has set up two such assemblages by the present lakeshore, where one would expect a good deal of disruption by the incessant waves; others are in quieter locations, such as by lagoons or on the banks of rivers. So far she has established more than twenty such 'camp-sites', and her intention is to document what happens to collections of stone tools in particular environments over a period of years. The data from this work will be of great assistance to archaeologists who need to reconstruct their sites.

Broken bones: hyaenas or hominids?

The lessons of modern stone knapping and the evidence from the careful reassembly of stone fragments should convince even the most sceptical of observers that the 'unpromising jumble' of stones on the ancient river bank at site 50 is indeed the litter left by early tool-making hominids. But what of the possibility that the bone fragments were the work of hyaenas, or some other carnivore, rather than of ancient hominids? Henry Bunn, who directs much of the excavation work at the site, has been examining fossil bones for a number of years now and, like Nick Toth, he has been performing experiments with modern material in order to answer this question.

Of the 2,000 pieces of fossil bone that have been uncovered from the site, 1,000 remain unidentified: these are mainly very small. From the rest it is possible to establish that parts of the following animals were buried at the site: a large species of hippopotamus, a pygmy hippopotamus, a zebra-like animal, a giraffe, a pig, an eland-sized antelope and a catfish. How do we know that these broken bones were the work of our ancestors? The answer is that a bone broken by hyaenas or by natural forces will not look exactly the same as a bone smashed open by human hands.

Take, for instance, the large leg bone of the eland-like animal at the site: it is smashed into at least ten pieces. The fragments show every sign of having being produced when the whole bone was placed on an anvil stone and hit a number of times along its length. Such an action yields characteristic 'percussion flakes', and this is what gave Henry Bunn an important clue. 'I realized that the humerus had been broken open in this way when I put together the evidence of my own experiments on bone-breaking and the experience I'd had when I was with the San in Botswana. The San smash open bones in this way so as to extract the marrow. It's a fair bet that what we see in the eland-like humerus, and in a number of other bones at the site, indicate that it was marrow the hominids were after.'

As it turned out, that smashed humerus bears yet more telltale signs of butchery. They are not easily seen, but in the correct light small cut marks stand out clearly. These marks were left by the sharp edge of a stone flake as it was used to slice the meat away from the bone. To show that an implement as rudimentary as a stone flake can make a significant mark on bone, Henry used one when there was a cow to be butchered. 'I found I was making cut marks on the humerus just like those on the fossil bone at site 50,' he says. 'Fresh bone is really very soft, especially at the ends.'

Henry Bunn's investigations, in combination with the efforts of the rest of the team, have demonstrated that these bones were collected at the river location known as site 50 through the action of hominids. Knowing this, one can safely guess that if it were possible to travel back to the moment when the hominid band was leaving its temporary home, one would almost certainly see a good deal more evidence of meat-eating than has been excavated. Much of that evidence was probably removed a short while afterwards by scavenging animals. Leave any fresh bone, or even a piece of dried hide, out at night in Africa, and the chances are that by next morning it will have been taken off and devoured by one scavenger or another.

Site 50, as it appears today, is probably a much diminished picture of what was once a camp-site littered with many bones. It is also likely that, just after the hominids had left, there would have been abundant evidence of plant foods, such as berries, succulent shoots, tubers and corms. As will be seen in the next chapter, people in tropical regions who make a living through hunting-and-gathering rely far more on plant foods than they do on meat, and this was almost certainly the case for our ancestors as well.

The use of a stone tool to slice meat from a bone leaves telltale cut marks, as seen on this antelope bone from Olduvai Gorge. Such marks clearly distinguish the remains of hominid meals from bones accumulated by hyaenas or other carnivores.

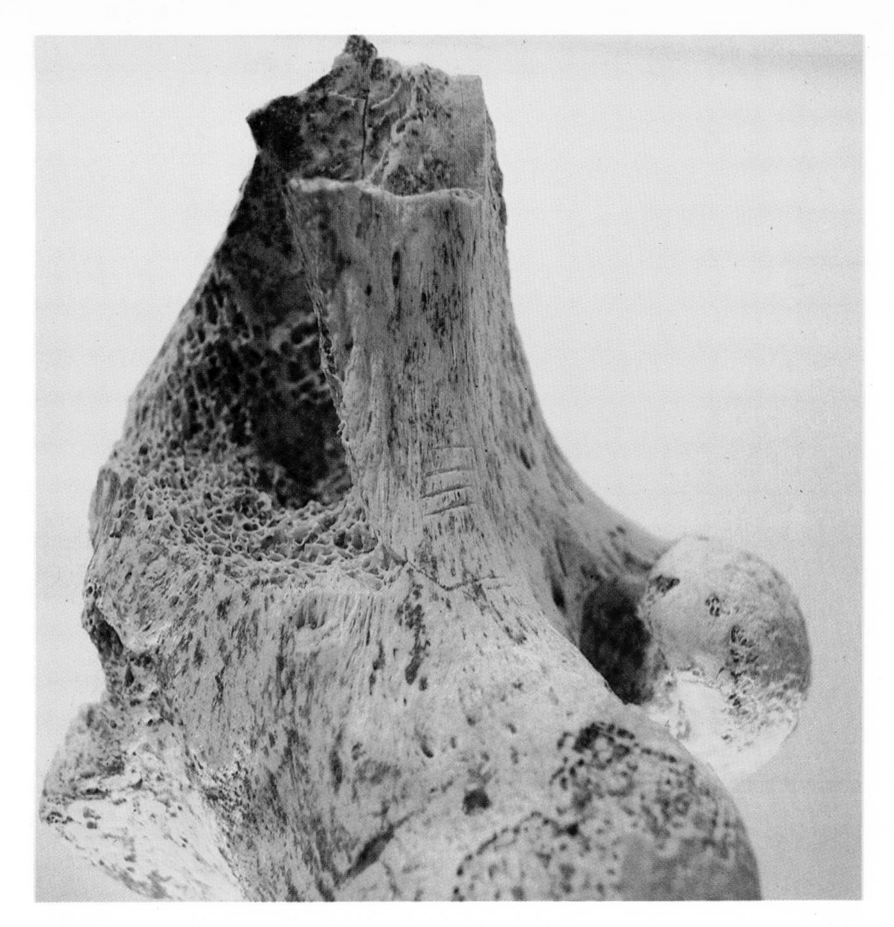

The lack of plant material in the fossil record has certainly led to an over-emphasis on meat-eating as a component of the early hominid's everyday life. Not until 40,000 years ago is there any really good evidence of the part played by plant foods in our ancestors' diets. So what chance is there for such evidence from a million-and-a-half years earlier? One possibility that is still in its early stages comes from the careful work of Larry Keeley at the University of Illinois.

For some years Larry Keeley has been making microscopic studies of the cutting edges of ancient stone tools, mostly flint tools from England and Belgium. He was interested in establishing whether it is possible to say what type of material a prehistoric implement had been used on. It was not an easy task: a number of researchers had tried previously, but with only qualified success. Nevertheless, Larry Keeley developed techniques for cleaning flint tools and viewing them under an optical microscope at 200 times magnification. He saw what no one else had seen before: different types of 'polish' on the flint surface that were characteristic for tools that had been used on meat, bone, hide, antler, wood or soft plant material.

Larry has now examined a series of flint tools, all approaching a quarter-of-a-million years in age, from three sites in England, and has managed to demonstrate signs of meat-cutting, wood-working and hide-working on some of them. One interesting discovery was that only a small proportion of the tools appeared to have been put to any use at all. Larry Keeley also successfully took up the challenge of Mark Newcomer, an expert stone-tool maker at London's Institute of Archaeology, to identify the materials on which he used a set of fifteen tools. Mark Newcomer had been more than a little sceptical about Larry Keeley's method, but he was convinced by the results. Larry Keeley has now begun to turn his powerful new technique onto the older stone implements of East Africa.

One problem in doing this is that most of the tools in East African sites are made from lava, and this is physically quite different from flint, which has a smooth, almost crystalline surface structure. The 'polish' from worked materials is almost certain to be different on the comparatively rough lava surface. Indeed, it is possible that no consistent polish will be produced at all. As yet, this remains to be seen. Meanwhile, Larry Keeley has begun to examine some of the site 50 tools which were made from chert, a more finely grained material than lava. In spite of problems with cleaning these tools, he has clear evidence of butchering usage on some small flakes. Further work will show whether the cutting of plant material has left any distinctive mark upon the flakes.

The bias against plant foods in the archaeological record is increased because collecting of plant foods requires almost no technology, and the few tools likely to be used in gathering plants are made from perishable materials. It is possible to collect large amounts of nutritious plant food with nothing more than a simple container such as a bark tray. Only when the search goes below ground, for roots and tubers, is a special tool, the digging stick, needed, and even this would not appear in the archaeological record, except possibly for traces of wood-working on stone flakes used to sharpen digging sticks. By contrast, to break into meat-eating in any significant way, hominids, with their modest teeth, needed to use sharp cutting edges. The resulting tools stand out like a beacon in an otherwise dim record of the past. But that beacon can lure unwary archaeologists into making ill-founded deductions. Glynn Isaac and his colleagues know this very well, and accept that the archaeological record may always present a biased image of the past.

Asked what has been achieved in nine months of digging at site 50, Glynn Isaac points out that 'the site has provided particularly clear evidence of some things that early hominids *were* doing: they repeatedly carried stones to certain favoured places and made simple sharp-edged implements from them. To those same places they seem to have carted parts of animal carcases. Once there they presumably ate the meat and they certainly broke the bones to get at the marrow. When people ask why the hominids did not eat their meat where they obtained it, I can point out a number of potential reasons. It is possible that they simply came to eat in the shade, but it seems even more likely that they carried food to special places like site 50 for social reasons – very particularly in order to feed youngsters, or even to feed their mates and relatives. Such food-sharing behaviours certainly became a universal part of the human pattern at some stage in evolution and many archaeologists are inclined to think it might have begun by the time site 50 formed. But this doesn't mean that the hominids who came to site 50 lived exactly like humans – for all we know they socialized there during the day and actually slept in trees. Equally, as yet we have little idea whether they got meat mainly by hunting or by scavenging; nor do we know what plant foods they may have eaten and whether these too were brought to the site. Only patient excavation of a whole series of such well preserved sites can help answer these questions.'

Sharing as a way of life

There is, therefore, a strong suggestion that the hominids of site 50 shared the food they had collected with other members of their group. To an

Overleaf: A reconstruction of the scene at site 50 when a band of hominids camped there one-and-a-half million years ago. There are no remains of the hominids themselves from which to identify them, but it is believed that they were *Homo erectus*.

anthropologist this is a very familiar pattern of behaviour: it is precisely what hunter-gatherer people, of whom there are a few remaining in the world, do every day of their lives. But to a non-human primate it is an extraordinary way of going about things.

Chimps and baboons are highly social, opportunistic omnivores. Both species eat a wide range of food, but chimps have a greater emphasis on fruit while baboons concentrate more on seeds, grasses and small roots – these differences reflect the type of terrain in which they live, chimps living in woodland and baboons in savanna. Meat is an occasional item, but it makes up no more than a twentieth of their diet.

Highly social though they may be in other respects, neither chimps nor baboons share their food in the way humans do. Several chimps may feed at the same fruit tree, but there is no sharing involved. The same is true for foraging baboons. Sharing is a totally unknown activity for baboons, while chimps will grudgingly engage in sharing only when an individual has caught an infant baboon or some other item of prey. Glynn Isaac prefers to call this sharing of meat 'tolerated scrounging', for other chimps must beg repeatedly to be given any of the prey.

The shift from individual feeding, which was probably practised by our very earliest ancestors, to collecting and sharing food at a home base was a profound alteration of lifestyle. What was behind such a dramatic revolution? Some people support a 'hunting' hypothesis, and others have proposed, as an alternative to this, the 'gathering' hypothesis. I, however, prefer Glynn Isaac's 'food-sharing' hypothesis.

A chimpanzee eating weaver ants. Chimpanzees, though highly social in other respects, never share their food, except on the rare occasions when an animal is captured. Even this sharing of meat is better described as 'tolerated scrounging'. Our earliest ancestors probably fed in this way, and the switch to co-operative food collection and sharing of food must have been a significant change of lifestyle.

The 'hunting' hypothesis has its basis in the notion of 'Man the Hunter', which anthropologist Sherwood Washburn did a great deal to develop. At a symposium in Princeton in 1956 he said: 'Hunting not only necessitated new activities and new kinds of co-operation but changed the role of the adult male in the group. Among the vegetarian primates, adult males do not share food. They take the best places for feeding and may even take food from less dominant animals. However, since sharing the kill is normal behaviour for many carnivores, economic responsibility of the adult males and the practice of sharing food in the group probably resulted from being carnivorous. The very same action that caused man to be feared by other animals led to food-sharing, more co-operation, and economic interdependence.'

The argument seems persuasive when one considers the example of the co-operative, social carnivores such as hunting dogs but only a few carnivores share their kill in this way. Indeed, many carnivores are solitary except in the breeding season. But ten years later, in 1968, Sherwood Washburn together with C S Lancaster pressed the point even harder. They wrote: 'To assert the biological unity of mankind is to affirm the importance of the hunting way of life. It is to claim that, however much the conditions and customs varied locally, the main selection pressures that forged the species were the same. The biology, psychology and customs that separate us from the apes – all these we owe to the hunters of time past.' In 1976 Robert Ardrey summed up the hunting hypothesis, in a book of the same name, with a characteristically dramatic statement: 'Man is man, and not a chimpanzee, because for millions upon millions of years we killed for a living.'

Anthropologist Sally Slocum rejects the hunting hypothesis: 'It leads to the conclusion that the basic human adaptation was the desire of males to hunt and kill. This not only gives too much importance to aggression, which is after all only one factor of human life, but it derives culture from killing. . . . Too much attention has been given to the skills required for hunting and too little to the skills required for gathering and the raising of dependent young.'

Sally Slocum points to the fact that among most contemporary hunter-gatherers, plant foods collected by the females provide the greater proportion of the daily diet. She suggests that the trigger that set hominids in this new direction was the ever-lengthening period of infant dependency: 'The mothers would begin to increase the scope of their gathering to provide food for their still dependent infants.' Adrienne Zihlman and Nancy Tanner have developed the gathering hypothesis further. Sharing of food began between mother and dependent infant, and these social and economic links persisted within what they call 'kin groups'. Males could be involved if they were kin, or, as the system evolved, if they too were providing; the male role was, however, somewhat peripheral. Adrienne Zihlman has recently written: 'The new and fundamental elements in human life included food-sharing as a matter of survival, regular sharing between mother and offspring and the expansion of the sharing networks to include adult females giving to adult males. . . . Females also shared with their male siblings. Later, these behaviours would be a basis for generalizing the sharing with adults outside the immediate kin group.'

The economically uncertain activity of hunting is seen as a relatively harmless pursuit that could be tolerated. 'So, in spite of time-consuming behaviours which frequently yield no food, such as hunting or obtaining raw materials from some distance away, individuals engaged in such activities, probably primarily males, could follow these pursuits because they were assured of a share of the food gathered by women with whom they had close ties.'

Nancy Tanner and Adrienne Zihlman turn the focus of the gathering hypothesis on the rise of technology: '*Tools* were used *not for hunting* large, swiftly moving, dangerous animals *but for gathering* plants, eggs, honey, termites, ants and probably small burrowing animals. Tools included sticks for digging or knocking down, rocks for cracking open nuts or fruits with tough rinds, and several types of containers. Sharp-edged rocks were perhaps also used for cutting some of the new savanna foods, especially roots and tubers, which were tough and fibrous but large enough to be divided. This is part of an overall shift, with hands and tools replacing some functions of teeth.' Both hypotheses have points to commend them, though the gathering hypothesis is considerably stronger than its hunting counterpart. It is highly probable that plant foods were indeed the major part of the early hominid economy, and unequivocal evidence for hunting as against scavenging carrion does not appear until relatively late in the fossil record, probably not earlier than half-a-million years ago.

The problem with both these proposals, however, is that they highlight one activity and one source of food as against another. The attraction of Glynn Isaac's food-sharing hypothesis is that it suggests an economy based on both meat and plant foods, the immediate benefit being that the hominids following this way of life would broaden the range of resources on which they depended for existence. Evolutionary success is principally about maintaining or improving access to food: expand your range of foods and you instantly obtain an advantage over your neighbours with a more limited diet.

As mentioned earlier, the technology needed to support the gathering

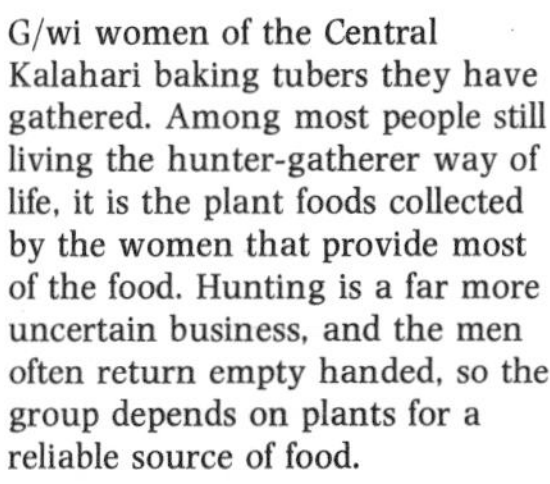
G/wi women of the Central Kalahari baking tubers they have gathered. Among most people still living the hunter-gatherer way of life, it is the plant foods collected by the women that provide most of the food. Hunting is a far more uncertain business, and the men often return empty handed, so the group depends on plants for a reliable source of food.

aspect of the mixed economy is very simple: some kind of container and possibly a digging stick. Indeed, anthropologist Richard Lee has suggested that the invention of some form of container was *the* technological innovation that led eventually to modern man. Given the knowledge that, struck in the appropriate manner, stones break and yield sharp cutting edges, early hominids engaged in gathering could then also capitalize on any dead animals they happened to find. This casual scavenging could rapidly have given way to a more active search for dead meat, and vultures circling in the sky may have been used as an indication of a potential meal. There is almost always some meat to be scavenged on the savanna which could have made a useful, if not always predictable, addition to the hominid diet. One advantage that meat undoubtedly has over plant foods is that it is a highly concentrated form of protein and fat. A lucky find could therefore have represented a food bonanza upon which to feast, leaving more time free in which to do other things: compare the herbivorous elephant's devotion of seventy-five per cent of its day to the business of eating with the carnivorous lion's fifteen per cent.

Hominids embarking on the mixed economy of collecting meat and plant foods would be trading some of their individual independence for greater economic security. Individuals would both contribute to and take from the group's collective efforts, with each individual doing much better than if he or she had attempted to collect food alone. Scavenging, combined with plant-food gathering as an insurance base, is a viable way of life.

The new economic order of the early hominids separated them from their ape-like cousins, not so much in what they ate, but rather the manner in which they ate it. Certainly, the hominids included more meat in their diet than their non-hominid relatives, but that difference was merely one of degree. The significant departure was the strategy of collecting food *to be eaten later*, and the consumption of food within a social network. An immediate consequence of such an arrangement would have been that social interaction, already well developed in higher primates, was enhanced still further.

An essential element of the food-sharing hypothesis as Glynn Isaac envisages it is a division of labour between males and females. The collection of meat, particularly when it involves active hunting, takes individuals farther afield than is demanded by the gathering of food plants. There is also the possibility of physical danger in the quest for meat. So it makes good sense for females, encumbered with young infants, to forage for plant foods, leaving the meat to the males. The contract of the mixed subsistence economy would, therefore, be essentially between the males and the females within the social group. This division of labour between the sexes is, in fact, seen in most modern hunter-gatherer communities.

I believe that the food-sharing hypothesis is a very strong candidate for explaining what set early hominids on the road to modern man. Not all the hominids survived to the present of course: *Australopithecus africanus* and *Australopithecus robustus* existed successfully for a couple of million years and then slipped into extinction. Only the *Homo* line continued. Does this mean that only early *Homo* adopted the food-sharing strategy? Did the australopithecines not make stone tools or take meat to a home base? Unfortunately, it is very difficult to see how archaeology can help in answering these important questions directly.

Once again it should be stressed that the evolutionary process is 'blind': it does not involve an objective, a specific goal of development on the part of the evolving species. Our *Homo* ancestors happened to fit into a particular ecological niche because they were endowed with the mental equipment suited to exploiting it. The niches occupied by *Australopithecus africanus* and *Australopithecus robustus* were undoubtedly different from each other and from that of *Homo*. The australopithecines should not be seen as failures in some grand evolutionary race to become human. They were very successful species pursuing stable lifestyles for several million years. The particulars of their way of life will be difficult to unravel from the tangle of evidence of hominid activities between four million and one million years ago. But one should view the australopithecines as creatures who, in times past, were well adapted to their environment.

However, the australopithecines did become extinct. Exactly why, we will probably never know. Possibly their lifestyles, though different from that of *Homo*, were similar enough for there to be some competition, which intensified as *Homo* became more and more successful. It may be, however, that competition came from a different direction: from the baboons. The demise of the australopithecines coincided with an increase of open-country baboons, animals that almost certainly sought out a similar ecological niche. Competition for food and space would have been inevitable. It is also possible that competition could have come both from *Homo* on the one hand and the baboons on the other.

Whatever the reason for their passing, the extinction of the australopithecines meant that there remained on earth just one hominid: *Homo*.

6 Life as a Hunter-gatherer

For at least two million years our ancestors followed a technologically simple but highly successful way of life. The initial strategy of the opportunistic scavenging of carrion combined with the organized collection of plant foods gradually evolved into a hunting-and-gathering way of life, the transition probably taking place somewhere between one million and half-a-million years ago. Not until relatively recently, between 20,000 and 10,000 years ago, did that long-established lifestyle begin to be replaced by systematic food production in the form of pastoralism or agriculture. The change came late in our history, but it developed with astonishing speed and is now virtually total. Only a handful of people, living in isolated parts of the globe, still subsist by the ancient hunter-gatherer way of life.

Hunting-and-gathering was a permanent and stable feature of our biological evolution through *Homo erectus* to early *Homo sapiens* and finally to modern man. Given the importance of hunting-and-gathering through the many thousands of generations of our forebears, it may well be that this way of life is an indelible part of what makes us human. Prehistorians like Glynn Isaac try to piece together scraps of evidence from the fossil record to discover what they can about early human behaviour, as described in the previous chapter. This kind of work has revealed a great deal, but, inevitably, it is limited in what it can expose of complex social behaviour. We know that until the advent of agriculture, humans made a living by gathering plant foods and hunting or scavenging meat, activities that centred on some form of home base. We can assume from this that there must have been some social organization, but the fossil record remains silent about what it was like to be a member of such a hunter-gatherer group. It indicates nothing of what was important to individuals in the group or what moral codes they adhered to, and it only hints at the skills required in order to survive. Of course, one can never know for certain the answers to these questions. But one can obtain some clues through the careful study of contemporary hunter-gatherers.

The image of hunter-gatherer societies

Nineteenth-century anthropologists viewed hunter-gatherers as fossilized societies, primitive savages who had somehow slipped unnoticed and unnoticing into the modern world. This is, of course, nonsense. The hunter-gatherers were as modern in biological terms as the explorers who 'discovered' them; they just happened to be sustaining themselves by an ancient method. Misconceptions about non-agricultural people abounded, often inspired by Thomas Hobbes's seventeenth-century notion of life in a state of nature: 'No arts; no letters; no society; and which is worst of all, continual fear and danger of violent death; and the life of man, solitary, poor, nasty, brutish, and short.'

A !Kung woman at her hut on the northern fringe of the Kalahari Desert. Many of the !Kung have become settled and have built permanent huts such as this, but some still live by hunting-and-gathering, a way of life very similar to that which our ancestors followed.

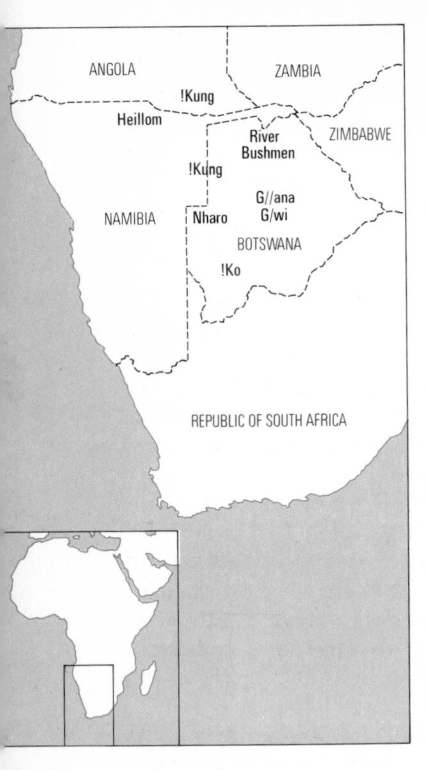

The !Kung are just one group of the San people, formerly called Bushmen, who live in and around the Kalahari Desert. The lifestyle of the San groups varies depending on the particular type of environment they inhabit.

During the past couple of decades, and particularly in the last ten years, the image of hunter-gatherers has undergone a transformation. Writing in his classic book *Stone Age Economics*, anthropologist Marshall Sahlins argued that in studying hunter-gatherers, Western anthropologists must not impose Western, that is, materialistic, ethics on their subjects. This goes farther than just overcoming the revulsion of nineteenth-century explorers at certain food items. (The consumption of large, juicy insect larvae, for example, was often assumed to be the desperate act of starving people, whereas the hunter-gatherers considered them to be great delicacies.) Marshall Sahlins refers to the different goals of the different societies: the pursuit of wealth, property and prestige in the one, and something totally different in the other. He even goes as far as to suggest that the hunting-and-gathering way of life is 'the original affluent society . . . in which all the people's wants are easily satisfied.' As it happens, the hunting-and-gathering economy is not an incessant search for food, as many anthropologists have supposed, but a system that allows a good deal more leisure than is possible in either agricultural or industrial society.

Marshall Sahlins's conclusion rests in part on an important study carried out on the !Kung San (formerly called Bushmen, a derogatory term coined by Dutch colonialists in southern Africa), who live close to the border between Botswana and Namibia, on the northern fringe of the Kalahari Desert. Since 1963 a number of researchers, principally based at Harvard University, have been analysing many aspects of the !Kung's hunter-gatherer way of life. Since the project began the inevitable march of 'progress' has impinged on the region and only a very small number of people in the area still hunt-and-gather for a living, the rest having been persuaded to settle down as agriculturalists. Some of the !Kung have even been recruited by the South African government as anti-terrorist trackers along the Botswana–Namibia border. The transition from a nomadic to a settled existence has in fact been highly instructive about the social components of each way of life, as will become apparent in Chapter 13.

Richard Lee, one of the principal investigators, recalls his motives for embarking on the study: 'I wanted to get away from the earlier misconceptions about hunter-gatherers. I wanted to find out what were the important elements of their way of life, without romanticizing them either in the Hobbesian "nasty, brutish, and short" manner or by putting them in a Garden of Eden.' Richard Lee is confident that his work is a legitimate way of gaining a glimpse of the past. 'The !Kung are a good model,' he claims, 'because, compared with prehistoric hunters and gatherers, they are living in a very marginal environment. Hunters of the past would have had the pick of rich resources and would undoubtedly have had an easier time of it than do the modern San.' The strength of the conclusions based on the study of the !Kung is that they are largely corroborated by other observations – some anecdotal, some scientific – on hunter-gatherers in many parts of the world. Through such peoples one can gain a valuable impression of the social and technical implications of the hunter-gatherer existence. One does not see exact replicas of our ancestors, but one can understand the principles that governed their lives.

Life on the edge of the Kalahari

To many people, the name Kalahari conjures images of a desert of unrelenting aridity. To the !Kung, however, it is home and has been for at least 10,000 years. It is a place where they make a very reasonable living in a manner that, until recently, had remained unchanged for millennia. Richard Lee describes the area in the following manner: 'The first impression of a traveller to this region is of an immense flatness where the sky dominates the landscape. The Aha Hills rise only a hundred metres above the surrounding plain, and from their top one sees what seem to be endless vistas of brush and savanna stretching to the horizon in every direction. . . . At several points in the landscape the sandy plain is broken by dry river courses. . . . They rarely hold water, perhaps twice in a decade, but when they do the flow can be considerable. . . . At night the stars overhead have an unbearable beauty, with the crystal clear high desert air and the central spine of the Milky Way galaxy arching overhead. It is with good reason that the !Kung name the Milky Way *!ku!ko!kumi*, the backbone of the sky.'

The desert winds have sculpted the sands into long, low red-topped dunes which run from east to west. Groves of mongongo nut trees cover many of the dune ridges, a feature of Kalahari life that is vital to the !Kung both for food and water. Gigantic baobab trees stand here and there, often the largest physical object within view. Everywhere is 'unbroken, unhumanized bush'.

Richard Lee chose an isolated group of !Kung to study, in an area he called the Dobe, which has nine water holes and is inhabited by about 450 people. He tackled and overcame the immense challenge of learning the !Kung language: 'The !Kung word can be described as an explosion of sound surrounded by a vowel. The bundle of clicks, fricatives and glottal stops that begins most words makes !Kung a difficult language to record, let alone to speak.' (The use of !,/, and other such marks in written !Kung represents the various clicks and explosive sounds.)

Once settled with his chosen group of !Kung, Richard Lee inevitably was exposed to the predominant feature of hunter-gatherer life: its mobility. During the wet summer season, from October to May, the small foraging bands erect modest temporary camps among the mongongo nut groves, moving on to new camps every few weeks. The band moves, not in a constant and desperate search for food, but because the longer people stay in one place the further they must walk each day in order to collect food. It is a question of convenience, not a flight from starvation.

The foraging bands at this time of year are small, having about six families in them. At least some of the families are likely to be related to each other, either by blood or by marriage. In any case, the !Kung create an extensive network of informal affiliations throughout their neighbouring bands. Giving gifts and receiving them (not trading) are essential elements in weaving together the social fabric of !Kung life, and relationships within and between bands are complex and close.

The numerical composition of a foraging band, roughly thirty people, has been called one of the 'magic numbers' of hunter-gatherer life. Throughout the world hunting-and-gathering people have as the core of their social and economic life a band of about this size. It appears to be the optimum combination of adults and children for exploiting the widespread plant and animal foods that hunter-gatherers live on: fewer than this and the social structure is weakened; more, and the work effort has to be increased in order to collect enough food for everyone. Incidentally, Glynn Isaac considers that the site 50 camp was probably occupied by around thirty people, although this is obviously a rough estimate. 'Most of the prehistoric living sites we find are of this rough order of size, until one approaches the Neolithic period,' he says. Only when the mode of production changes from the basic hunting-and-gathering system to more settled agriculture do groups larger than thirty become viable over long periods.

When the dry winter months come, the !Kung congregate around permanent water holes in concentrations of a hundred or more people. This

The nomadic !Kung congregate around permanent water holes during the dry season in groups of a hundred or more. This is an opportunity for a great deal of social activity, including story-telling and large-scale trance-dancing. The dances are part of a healing ceremony and often go on all night: as the sun rises the people begin to dance with renewed vigour.

'public' phase of their life is very important. It is the time of intense socializing, large-scale trance-dancing and curing, initiations, story-telling, exchange of gifts and marriage-brokering. If one sees the dispersed summer camps as being connected by an invisible network of kinship, friendship and material obligation, then the winter is the time when the net is pulled together, bonds are strengthened and new alliances made.

This highly valued public phase of !Kung life is not without its drawbacks, however. The unusual concentration of people inevitably means that people have to do more work: they have to travel farther to collect plant food and to find animals to hunt. And with the high-density living there often comes personal conflict. As soon as the rains begin, people once again disperse in small bands to the mongongo groves. The bands are, however, not necessarily of the same composition as those of the previous summer: the public phase is an opportunity for people to join with others with whom they would prefer to live and for tensions and conflicts to be resolved by the splitting up of some bands.

Public-phase camps and the fission and fusion of bands are very common among foraging people. For instance, the G/wi San of the much more arid Central Kalahari live in small bands for about 300 days of the year without any standing water. They survive for most of this time on moisture obtained from melons and tubers. During the summer rains, however, they congregate around the few meagre and ephemeral water holes. The ostensible reason for the coming together is to take advantage of the standing water, which is briefly available. But the opportunity for intense socializing is not lost, and one suspects that the prospect of mixing with relatively large numbers of people is anticipated with even more relish than is the chance of drinking from a pool.

During the wet season the !Kung live in small bands and move on every few weeks to a new base camp. In this way they avoid exhausting the food resources within easy reach of the camp.

Hunter-gatherer peoples frequently give elaborate reasons why they come together in large groups and then break up again, but the real motive seems to be a strong need for formal and informal socializing within a large number of relatives, friends and affiliates. While non-human primates frequently live in social groups of about thirty individuals, they do not usually come together in the larger groups that are such a feature of the human hunter-gatherer way of life.

Hobbes's view that non-agricultural people have 'no society' and are 'solitary' could hardly be more wrong. To be a hunter-gatherer is to experience a life that is intensely social. As for having 'no arts' and 'no letters', it is true that foraging people possess very little in the form of material culture,

The G/wi people depend almost entirely on melons and tubers for their moisture during the dry season. Only in the wet season are there any water holes for them to drink from.

but this is simply a consequence of the need for mobility. When the !Kung move from camp to camp they, like other hunter-gatherers, take all their worldly goods with them: this usually amounts to a total of 12 kilograms (26 pounds) in weight, just over half the normal baggage allowance on most airlines. There is an inescapable conflict between mobility and material culture, and so the !Kung carry their culture in their heads, not on their backs. Their songs, dances and stories form a culture as rich as that of any people.

In spite of the apparently inhospitable aspect of the northern Kalahari, it does in fact support a large number of wild animals. Richard Lee explains: 'To give some idea of abundance, fresh warthog, steenbok and duiker tracks can be seen every day of the year. Kudu, wildebeest and gemsbok tracks might be seen several times a week. Tracks of giraffe, eland, hartebeest and roan antelope would be seen perhaps once a month or once in two months; buffalo have been sighted only about a dozen times in ten years, zebra perhaps three times and impala only once in the same period.' The natural array of plant foods is impressive too. 'The !Kung are superb botanists and naturalists with an intimate knowledge of their environment,' Richard Lee says. 'Over 200 species of plant are known and named by them, and of these a surprisingly high proportion is considered by the !Kung to be edible.'

A G/wi boy with a harp. The mobility of the hunter-gatherer way of life does not allow for the accumulation of material goods, and their culture consists largely of songs, dances and stories. It involves few possessions, apart from musical instruments and the beads they make from ostrich shells.

The role of women

Referring specifically to the !Kung, Richard Lee gave a paper at a symposium in 1966 entitled *What hunters do for a living*. This revealed that, contrary to popular conception, meat constitutes only thirty to forty per cent of the diet. Moreover, in studies of other hunter-gatherers in similar latitudes, the same

sort of figure cropped up again and again. The conclusion was inescapable: plant foods are the staple of hunter-gatherer life. Only in the higher latitudes, where the changing seasons make plant foods an unreliable resource, do hunter-gatherers turn to fish or meat for the bulk of their diet.

In virtually all foraging people, as with the !Kung, it is the women who do most of the gathering of plant foods while the men do most of the hunting. There are exceptions, of course, such as the Agta people of the Philippines where the women share the hunting with the men. But in the vast majority of cases, there is a sexual division of labour. The most obvious reason for this arrangement is the incompatibility between the demands of the hunt – the long distances travelled while tracking prey and the quiet and stealth that is critical during the final stalk – and the problems involved in carrying weighty and noisy infants.

!Kung women give birth once every three to four years. Once again, the birth interval of roughly four years is a worldwide phenomenon among hunter-gatherer peoples, and it appears to be a biological response to the physical demands of mobility. Very young children must be carried while gathering and on migrations from an old camp to a new location. Transporting two children *and* gathering food would be extremely arduous. Richard Lee has calculated the amount of work which would be involved in carrying infant- and food-loads if the birth intervals were one year, two years, three years and so on. At the shorter intervals the work load is enormous and decreases rapidly when the birth interval is about four years, but does not drop significantly as the birth interval gets longer than four years. The world's hunter-gatherers therefore appear to have hit upon the optimum spacing entirely independently of each other. Only with a sedentary existence, such as in an agricultural economy, can women have babies more frequently without imposing on themselves an enormous carrying burden.

Why some kind of crèche system is not more common among foraging people is a mystery, but it may be because the mother's presence is necessary for suckling, which, with !Kung mothers, goes on for three or four years, often long after milk has ceased to flow. This extended suckling may be a physiological mechanism for preventing ovulation and therefore reducing the chances of another pregnancy. On the other hand, a crèche would make long birth spacing less necessary and eliminate the need for prolonged suckling.

The consequence of the sexual division of labour and the long birth interval is that food gathering is a highly social activity involving several mothers and their young, whereas hunting is a much more solitary affair, usually undertaken by a pair of men, possibly with an 'apprentice' adolescent. The economic differences between hunting and gathering are also profound. A woman can gather in one day enough food for her family for three days, and she seldom fails. A man may bring down a large animal that will feed the band for several weeks, or he may come home with only a small spring hare. Often he returns empty-handed.

According to Lee's calculations, !Kung men work just over twenty-one hours a week in contributing meat to the camp whereas women spend slightly more than twelve hours a week supplying plant foods which constitute about seventy per cent of the diet. When all other forms of work – including making tools and housework, but excluding child care – are added together, a man's week is over forty-four hours and a woman's forty hours. But as women do most of the child care, their total work load is greater than the men's. Richard Lee considers that the women do not feel themselves exploited: they have economic prestige and political power, a situation denied to many women in the 'civilized' world.

!Kung children continue suckling until they are three or four years old. This may help to prevent the mother from becoming pregnant again during this period. By the time she does have another baby the first child is old enough to walk on food-gathering trips and when the band moves on to a new camp.

Two men of the Nharo San hunting.

The ethics of hunter-gatherers

Anyone going into a !Kung camp and expecting to find a cache of food is in for a surprise. Hunter-gatherers simply do not lay up stocks against future shortages. Such an 'extraordinary' attitude provoked nineteenth-century anthropologists into commenting that hunter-gatherers behaved 'as if the game were locked up in a stable', and that they had 'not the slightest thought of, or care for, what the morrow may bring forth.' Rodney Needham, writing in 1954, said that foraging people have 'a confidence in the capacity of the environment to support them, and in their own ability to extract their livelihood from it.'

Food storage would run counter to the !Kung's habit of sharing food, particularly meat. Perhaps because it is a relatively rare commodity, meat is highly prized by the !Kung, as it is among most hunter-gatherers. When an animal is killed, the hunter (or rather the person whose arrow struck the prey – and that is not always the same person as he who shot the arrow) initiates an elaborate process of sharing the raw meat. The sharing runs

along lines of kinship, alliances and obligations. Lorna Marshall, a pioneer in !Kung studies, once witnessed the butchering of an eland, the largest of the African antelopes, and she counted sixty acts of meat distribution within a short time of the initial sharing. The network of sharing and obligation is very important among the !Kung. Richard Lee emphasizes the point strongly: 'Sharing deeply pervades the behaviour and values of !Kung foragers, within the family and between families, and it is extended to the boundaries of the social universe. Just as the principle of profit and rationality is central to the capitalist ethic, so is sharing central to the conduct of social life in foraging societies.'

This ethic is not confined to the !Kung: it is a feature of hunter-gatherers in general. Such behaviour, however, is not automatic; like most of human behaviour, it has to be taught from childhood. 'Every human infant is born with the capacity to share and the capacity to be selfish,' Richard Lee says. That which is nurtured and developed is that which each individual society regards as most valuable.

In the same vein as the sharing ethic comes a surprising degree of egalitarianism. The !Kung have no chiefs and no leaders. Problems in their society are mostly solved long before they mature into anything that threatens social harmony. Although the !Kung are very thinly distributed overall – occupying on average about 4 square kilometres (1.5 square miles) per person – their camps, by contrast, are an intense compression of humanity. People's conversations are common property, and disputes are readily defused through communal bantering. No one gives orders or takes them. Richard Lee once asked /Twi !gum whether the !Kung have headmen. 'Of course we have headmen,' he replied, much to Richard Lee's surprise. 'In fact, we are all headmen; each one of us is a headman over himself!' /Twi !gum considered the question and his witty answer to be a great joke.

The stress on equality demands that certain rituals are observed when a successful hunter returns to camp. The object of these rituals is to play down the event so as to discourage arrogance and conceit. 'The correct demeanour for the successful hunter,' explains Richard Lee, 'is modesty and understatement.' A !Kung man, /Gaugo, described it this way: 'Say that a man has been hunting. He must not come home and announce like a braggart, "I have killed a big one in the bush!" He must first sit down in silence until I or someone else comes up to his fire and asks, "What did you see today?" He replies quietly, "Ah, I'm no good for hunting. I saw nothing at all . . . maybe just a tiny one." Then I smile to myself because I now know he has killed something big.' The bigger the kill, the more it is played down.

'The theme of modesty is continued when the butchering and carrying party goes to fetch the kill the following day,' Richard Lee explains. The helpers joke, complaining that surely the hunter did not need so many people to carry such a puny kill. And the hunter will agree, suggesting that they just cut out the liver and go and look for something more worthwhile. The jesting and understatement is strictly followed, again not just by the !Kung but by many foraging people, and the result is that although some men are undoubtedly more proficient hunters than others, no one accrues unusual prestige or status because of his talents.

When one examines the technology of the hunter, and that of the other aspects of !Kung life, one is very impressed not by its complexity or sophistication but rather by its simplicity. A club and bows-and-arrows, the arrows tipped with insect-larvae poison, are the hunter's prime equipment, though they also use hooks for retrieving spring hares from their burrows, snares for entrapping small animals, and net-bags made from animal sinews for carrying various items. For plant-food gathering, the technical array is even more modest: a digging stick and some kind of container. The !Kung women use a kaross, which is made from antelope hide, in which they can transport the nuts, fruits, roots and berries, together with their infant. Back at camp, crude stones are sufficient for the initial cracking open of the plentiful and valued mongongo nut. A wooden mortar and pestle are all that is required for pounding the inner nut meat, so as to render it digestible to the young and the old, or to mix it with other foods. And as water, or rather the lack of it, dominates !Kung life, containers, such as ostrich eggshells or bags made from animal stomachs, are also important.

By contrast with the simple equipment used in hunting-and-gathering, the *skill* demanded is prodigious. Hunters must be able to identify an animal from its tracks, know how old it is, how long ago it passed by, whether it was running or idling, and whether it is injured or healthy. Once within striking distance of the prey, considerable cunning is needed to approach close enough for a telling shot with the diminutive arrows. Only the smallest antelope is knocked down by such a shot and most prey has to be tracked for several hours, sometimes days, before the poison kills it.

The !Kung are finely tuned to their environment, reading it closely and more thoroughly than people from a non-foraging society could ever comprehend. Two ethologists, Nicholas Blurton-Jones and Melvin Konner, once talked with the !Kung hunters to see how much of the animal world they understood. They were astonished at the !Kung knowledge: 'Some !Kung observations which we refused to believe were later proved correct when checked with ethologists who have worked in Africa,' they admitted.

Gathering also requires a great deal of skill. Patricia Draper, who has made a special study of women in !Kung society, puts it this way: 'The !Kung economy looks simple in comparison with other more diversified economies with greater division of labour, but from the point of view of the individual actor, subsistence is quite complex. For example, although it is simple enough to pick up nuts or melons once one is standing where they are found, it requires enough strength to walk 16 kilometres [10 miles] or more per day carrying a full day's harvest and perhaps a child. A woman needs to know where various foodstuffs are to be found, in what season they are edible, and how to keep oriented in the bush. !Kung women, like their men, pay close attention to animal tracks as they pass through the bush; and they tell the men about recent game movements when they return home in the evening.'

It is clear when looking at a wide range of hunter-gatherers that there are significant similarities in the way they run their lives. These similarities seem to be imposed by their way of life, and they are therefore important in forming an impression of the general character of our ancestors' lives. We can make the following general statements about hunter-gatherers:

The main social and economic focus of the hunter-gatherer existence is the home base, probably occupied by about six families. The main social consequence of the dual pursuit of meat and plant foods is a sexual division of labour, with the males doing most of the hunting and the females most of the gathering. The quest for meat is surprisingly unrewarding, and unless plant foods are too seasonal to act as a staple they will provide the largest proportion of the diet. The practice of bringing plant and animal foods back to a home base where they are shared with all members of the band demands a highly developed sense of co-operation and equality. The system also allows for an unusually large amount of spare time which people divide between visiting relatives and friends in nearby bands and entertaining visitors at their own camp. Indeed, the degree of socialization is intense and reaches a pitch when, for a short while, bands coalesce into larger groups. Above all, foraging people deploy tremendous skill and only minimal technology in exploiting their environment.

As Richard Lee says, 'We mustn't imagine that this is the exact way in which our ancestors lived. But I believe that what we see in the !Kung and other foraging people are patterns of behaviour that were crucial to early human development.' Of the several types of hominid that were living two to three million years ago, one of them – the line that eventually led to us – broadened its economic base by sharing food and including more meat in its diet. The development of a hunting-and-gathering economy was a potent force in what made us human.

7 New Horizons

I generally fly the two-hour-and-forty-minute trip from Nairobi to Koobi Fora every couple of weeks during the period from June to December. Kamoya Kimeu and I use the visits to discuss the progress of the fossil-hunting, and we spend some time examining closely the more promising fossils that have been discovered. Usually the visits are quite routine, with Kamoya's team continuing to accumulate a fine collection of fossils from the lakeside's ancient deposits. When I flew to the camp at the beginning of September in 1978, however, events proved rather more exciting: by good fortune I came across a magnificent cranium of an important human ancestor.

Kamoya and I were out on foot relocating a fossil site his men had discovered a week or so earlier at Ileret, some 70 kilometres (45 miles) north of the main camp. While we walked I was talking to Maundu, one of Kamoya's assistants, and he was reporting the news of the previous two weeks' work. As usual, we were all keeping a sharp look-out for signs of any interesting fossils as we went. Suddenly my eye was drawn to a small scatter of bone fragments protruding less than two centimetres through the grey sand. Maundu had also noticed the fragments and we both stopped and carefully knelt to inspect them more closely. We quickly agreed that they appeared to be some of the smaller bones from the underside of a hominid skull. Such finds are tremendously exciting and we immediately began to speculate about the chances of there being a whole intact cranium buried just beneath the surface. The answer had to wait until we could make a proper excavation the following day.

We returned the next day at dawn and soon revealed a problem I had feared: the bone was extremely fragile and as Meave and I gradually exposed more and more of the skull we had to apply a hardening solution of plastic to strengthen it. By 10.00 a.m. the sun was uncomfortably hot and we set up a tarpaulin shelter to provide shade in which to continue. But with the upturned cranium half-exposed, and the chance that it was intact now looking increasingly good, the noonday sun defeated us and we retreated to camp until the late afternoon. Once back to the excavation I soon completed the process of scraping away the surrounding matrix and applying the hardening solution. By 5.00 p.m. I was able to lift from the ground the delicate but intact cranium of a hominid who had lived and died close to the lakeshore almost one-and-a-half million years ago. Registered in the museum as KNM-ER 3883, the fossil is a fine example of *Homo erectus*, the hominid that immediately preceded *Homo sapiens*.

From other fossil material that has been unearthed, it is clear that the skeleton of *Homo erectus* was essentially modern. A little stockier than the average human today, perhaps, but not all that different. The head and face, however, were still 'primitive': the forehead sloped backwards and was

Homo erectus lived from about one-and-a-half million years ago to 300,000 years ago. These people were stockily built, but otherwise had a physique very like that of modern humans. Their facial features, on the other hand, were quite distinctive. Three *Homo erectus* men are represented here by modern people wearing masks designed on the basis of scientific measurements of *Homo erectus* skulls.

mounted with prominent brow-ridges, while the brain, though larger than that of *Homo habilis*, was only seventy per cent of the size of a *Homo sapiens* brain. The face protruded less than in *Homo habilis*, but it was not as flat or 'tucked in' as in *Homo sapiens*. The chin that is so characteristic of modern humans was present but poorly developed.

Palaeoanthropologists view the anatomical characteristics of *Homo erectus* as being distinctive enough to deserve recognition as a new stage in human evolution, an advance on *Homo habilis*. What is most striking about *Homo erectus*, however, is not the development of new anatomical features but the changes in behaviour. Through the development of the food-sharing, hunter-gatherer way of life and a sharpened intelligence, *Homo erectus* expanded into territories where no advanced hominid had lived before. Around a million or more years ago, some groups of this hominid moved into Europe and Asia. With this move, our ancestors changed from being exclusively tropical creatures, and learned to cope with the fluctuations in food availability that go with the changing seasons of temperate regions.

Homo erectus is first recognized in the fossil record from about one-and-a-half million years ago, and continued until about 300,000 years ago when *Homo sapiens* began to emerge. *Homo erectus* spread across the continents of Africa, Asia and Europe, though not to the colder northern extremes of Eurasia and not into America or Australia. It was the time when hunting-and-gathering became firmly established, with active hunting as opposed to opportunistic scavenging becoming more and more important. Large-scale hunting first developed in this period, we see the first signs of the systematic and controlled use of fire, and there are indications of ritual in the lives of these hominids. Stone-tool manufacture became controlled and patterned. New challenges were being faced and overcome, and new life-styles developed. The age of *Homo erectus* was clearly an important phase in human evolution, and it forms the subject of this chapter.

The search for the 'missing link'

Homo erectus first became known to palaeoanthropology through the determination of a remarkable Dutchman, Eugene Dubois. Born in 1858, Dubois was an anatomist and lectured at the University of Amsterdam. He developed a consuming interest in the nature of man's forebears, and was greatly influenced by the writings of Alfred Russel Wallace, the co-proposer with Darwin of the theory of evolution by natural selection. Wallace had lived in the Malay archipelago for some years and came to believe that this was probably where human ancestors had lived: it was, after all, the home of the orang-utan, the 'man of the forest'. For no good reason beyond a deep inner conviction that this was the part of the world in which to look, Eugene Dubois determined to go to Southeast Asia and find the 'missing link', the creature that would supposedly link man directly with the apes in a simple evolutionary sequence.

In 1887, accompanied by his wife and their newborn child, Dubois set sail for Sumatra (now part of Indonesia) on a venture that was universally thought to be complete madness. The focus of the world of palaeoanthropology at that time was Europe and, according to the leading academics of the time, there was simply no point in looking elsewhere. Dubois was determined to prove them wrong, and he did.

As no one would sponsor his scientific expedition, he had to support himself by working as a doctor for the Dutch East India Army, a post that left him ample spare time in which to pursue his searches. For two years he scoured the limestone caves and other promising locations on the island,

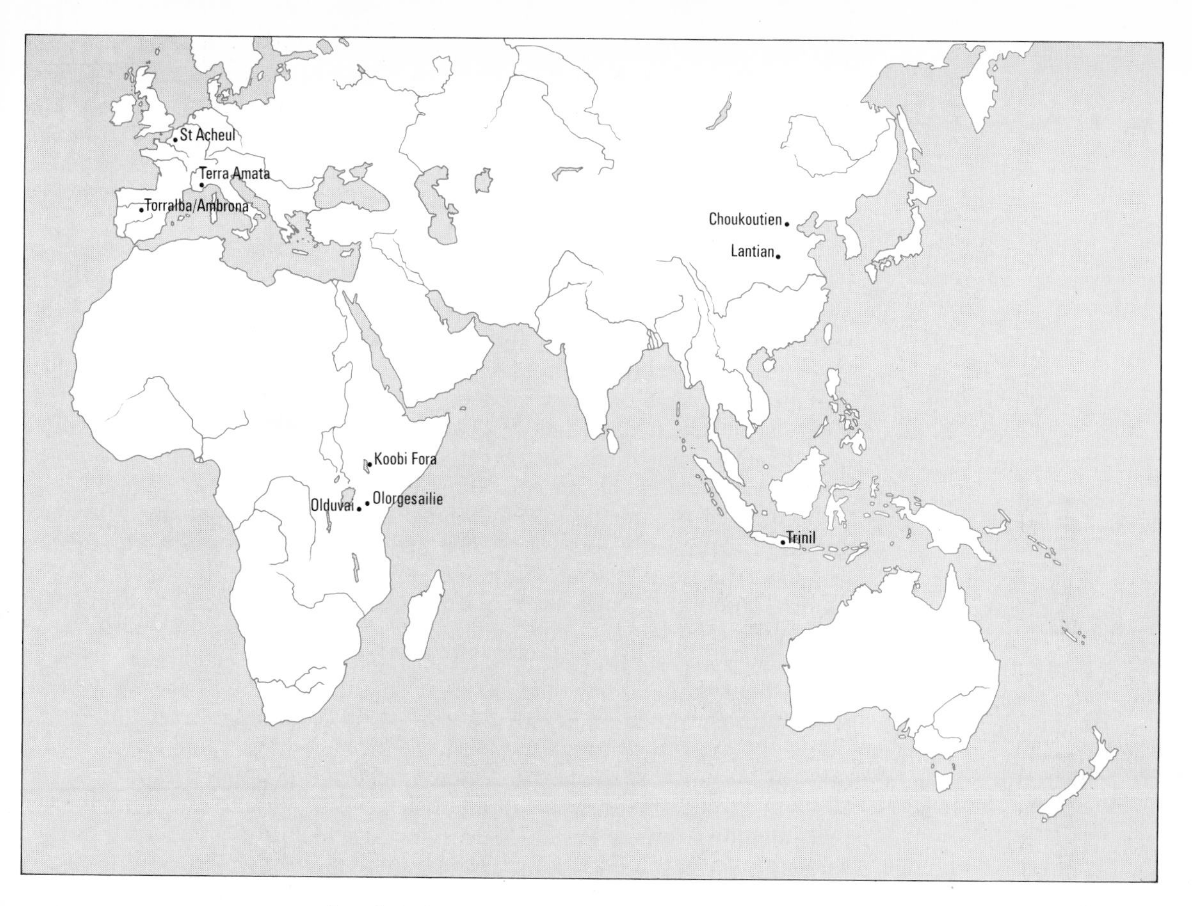

The major sites where *Homo erectus* remains have been found. This hominid was the first to migrate out of Africa and survive in the colder climates of the north.

but failed to find what he was looking for. Then, following a bout of malaria, he was transferred to nearby Java where he had even more time available for his quest. By now the Dutch government was showing some interest in his fossil-hunting activities and even gave him some support.

One interesting site that Dubois located was a 15-metre (50-foot) high embankment along the Solo River near the village of Trinil. River deposits, volcanic debris and sandstone layered the embankment, and the promise of good fossil material seemed strong. His workers began digging in August 1891 and within a month had recovered a curious ape-like tooth. Then came the discovery the Dutchman had always believed was inevitable: the skullcap of a human ancestor. The brain case was thicker than in modern humans and it bore prominent brow-ridges, features that Dubois interpreted as belonging to his 'missing link'. It was in fact the cranium of a specimen of *Homo erectus*.

The river then flooded with seasonal rains, and further excavations had to wait until the following year. When the team could continue their work, they unearthed a number of specimens of fossil pig, primitive elephant, rhinoceros, deer, tiger, hyaena and many other species. Then, ten months after the skullcap had been found, a fossil thigh bone of *Homo erectus* was discovered. These two important hominid fossils had been lying about 15 metres (50 feet) apart in the deposits. Dubois was ecstatic.

The thigh bone told him that the 'missing link' clearly walked upright, and he named the creature *Pithecanthropus erectus*, the 'erect-walking ape-man'. (The term *Pithecanthropus* had been proposed some years earlier by the German scientist Ernst Haeckel, who had drawn up a chart of human ancestry, showing a hypothetical predecessor of modern man, which he named *Pithecanthropus alalus*. The name *alalus* referred to the supposed absence of spoken language in this 'ancestor'. Naming a yet-to-be-discovered species in this way is rather risky and Dubois's discovery helped to rescue Haeckel from an embarrassing situation.) The name was changed to *Homo erectus* in the 1950s, following a major review of the evidence by Ernst Mayr.

Eugene Dubois spent much of the rest of his life attempting to convince the scientific world that what he had discovered was of importance, but he met with considerable scepticism and died a bitter man. Since then, however, four important fossil localities have been discovered on the island, and they were excavated in the 1930s and from the 1960s through to the present. They have yielded a number of interesting fossil crania and parts of lower jaws. Dubois has been vindicated in his belief that a form of primitive man lived in Southeast Asia, but there is one major problem with fossils from Indonesia: how old are they?

Because of the geological conditions under which the fossils were preserved, it is impossible as yet to measure their age with much confidence. As Garnis Curtis, a leading expert on dating, recently put it: 'Indonesia is a big question mark'. Dates ranging from as much as two million years down to as little as 700,000 years have been variously proposed for the fossil sites. If it turned out that the Indonesian *Homo erectus* lived at the earlier end of this time range, this would have important implications for our view of when human ancestors began to move out of Africa. An early date would also lend some support to those people who consider that *Homo erectus* evolved in Asia rather than in Africa. I believe this latter notion to be unlikely as, so far at least, none of the earlier, pre-*Homo erectus* hominids have been identified outside Africa. There is no knowledge of an ancestral stage, such as *Homo habilis*, from which *Homo erectus* could have evolved in Asia.

The migration from Africa

Since Eugene Dubois's pioneering work at the end of the last century, specimens of *Homo erectus* have been discovered in many parts of Europe, Asia and Africa. The move out of Africa was a major step and must have entailed radical changes in lifestyle. The tropics offer considerable security of food resources, with ripe fruits available throughout the year and meat of some kind available for most of the time. Not so in temperate regions, where the changing seasons bring succulent new plant growth in the spring, maturing plants during the summer, abundant nuts and fruits in the autumn, but almost nothing in the barren months of winter. The challenge of temperate regions was to adapt from having both plant and animal foods constantly available to a strategy that made use of different food at different times of the year. Clearly, *Homo erectus* was able to overcome what previously had been an ecological barrier to hominid migration.

The forebears of *Homo erectus* must have been reasonably intelligent creatures for their brains were almost twice as large as those of the australopithecines. They could manufacture simple stone tools, and they probably included much more meat in their diet than any other primate. Their food-sharing economy must have demanded much more complex social interactions than had ever existed previously. Intellect and co-operative food collection were the key factors in their success. Given time, these hominids evolved into *Homo erectus* whose substantially larger brain allowed it to operate the food-sharing economy even more successfully and develop it to new levels.

Homo erectus was able to exploit more resources in the environment than ever before. Alan Walker's studies of tooth wear (p. 74) show that one important breakthrough was the discovery of roots, bulbs and tubers. These can constitute a secure supply of food in an otherwise barren terrain, and may also be a welcome source of water, as the G/wi San of the Kalahari Desert know very well. A simple digging stick, and the knowledge of where to look, could immediately open up new possibilities to an intelligent creature.

A greater organizational ability, combined with keener observation of the habits of prey animals, would have produced new hunting skills. Instead of just waiting for the chance find of a recently dead animal, *Homo erectus* would also have deliberately sought out the young, weak and vulnerable animals in grazing herds, killing them by stealth and cunning. The extraordinary ability of modern humans to throw objects with force and with astonishing accuracy, even at moving targets, is interesting in this context. The brain mechanisms that underlie this unique accomplishment may have their evolutionary roots in the hunting activities of *Homo erectus*.

The increase in meat-eating had its hazards, however. The evidence for this comes from an interesting fossil specimen, found a couple of years ago at Koobi Fora. Several hundred pieces of a one-and-a-half million year old *Homo erectus* were discovered – part of the skull, the lower jaw, some vertebrae, ribs, leg bones and much of the pelvis – making up the most complete hominid skeleton so far unearthed from that time range. On the leg bones were intriguing signs of some kind of disease: the shafts appeared patchily encrusted with new bone.

The mystery was solved when a pathologist saw the diseased bone under a microscope and immediately recognized the condition as 'hypervitaminosis A' that is, a poisonous overdose of vitamin A. The disease is rare, but one of the most common ways of contracting it is to eat large quantities of raw liver, which is especially rich in this vitamin. When meat-eating was a novel pursuit for our ancestors mistakes of this kind must have happened occasionally, until a lore built up of what was good to eat and what should be avoided.

If one were to collect all the specimens of *Homo erectus* crania so far discovered, the anatomical similarities would be obvious: the large brain

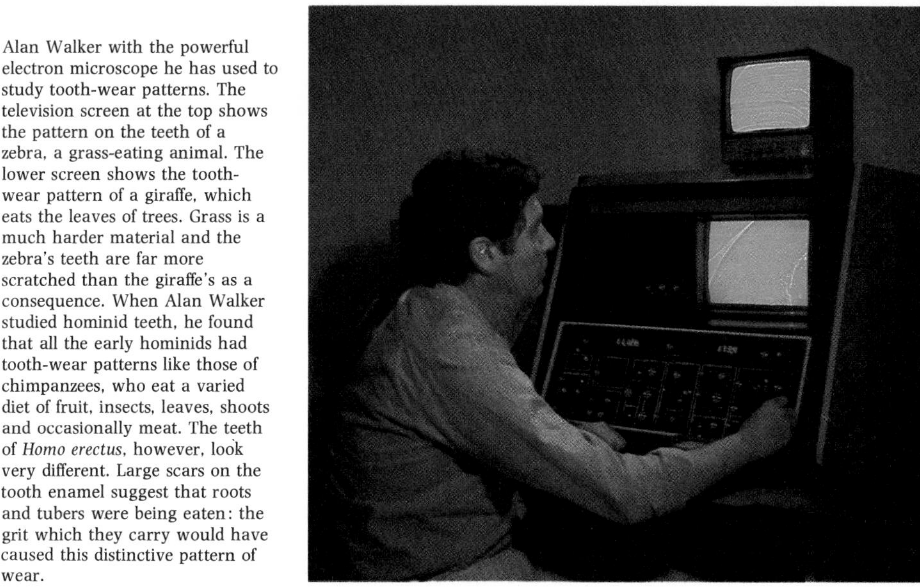

Alan Walker with the powerful electron microscope he has used to study tooth-wear patterns. The television screen at the top shows the pattern on the teeth of a zebra, a grass-eating animal. The lower screen shows the tooth-wear pattern of a giraffe, which eats the leaves of trees. Grass is a much harder material and the zebra's teeth are far more scratched than the giraffe's as a consequence. When Alan Walker studied hominid teeth, he found that all the early hominids had tooth-wear patterns like those of chimpanzees, who eat a varied diet of fruit, insects, leaves, shoots and occasionally meat. The teeth of *Homo erectus*, however, look very different. Large scars on the tooth enamel suggest that roots and tubers were being eaten: the grit which they carry would have caused this distinctive pattern of wear.

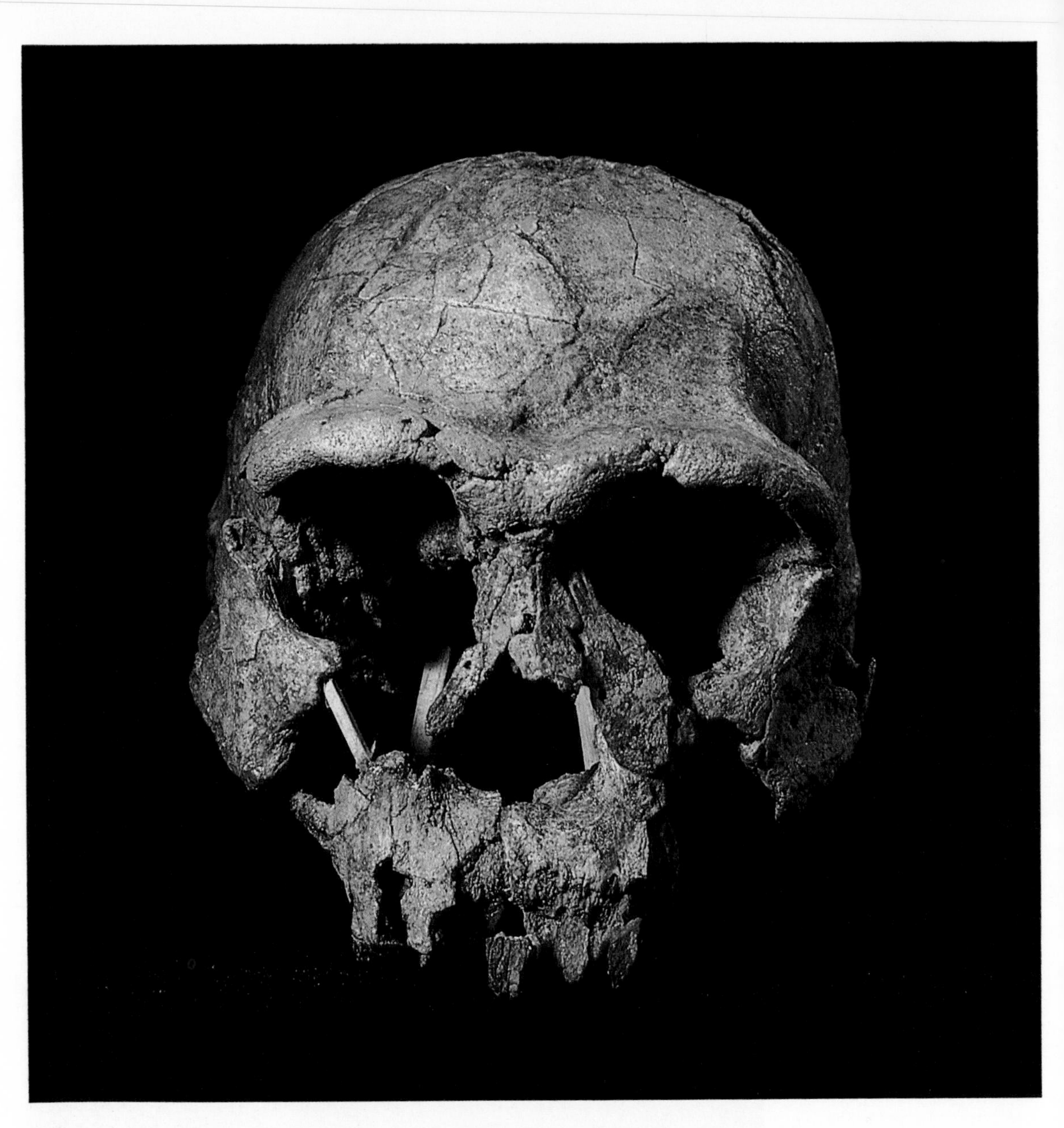

A *Homo erectus* cranium found at Koobi Fora in 1975.

case, the prominent brow-ridges, the conformation of the face and the thickness of the skull bone. One would also see differences: in the shape of the skull, in the degree of protrusion of the face, the robustness of the brows and so on. These differences are probably no more pronounced than we see today between the separate geographical races of modern humans. Such biological variation arises when populations are geographically separated from each other for significant lengths of time.

There is one biological factor that we can be certain about in those populations of *Homo erectus* that migrated out of the tropics and into the temperate regions of the world: their skin was dark in colour. The dark skin pigmentation of people living near the equator is essential protection against the harmful ultraviolet rays of the sun. Our more distant ancestors must have had a thick covering of body hair like that of gorillas and chimpanzees. It is an open question why this was reduced, to fine, short body hair, but it probably came about at the same time as our sweat glands developed and may have been a response to hunting on the open plains, where keeping cool became more of a problem than it had been for our ancestors.

Once this covering of hair disappeared, a dark-coloured skin was a biological necessity. But while strong ultraviolet rays are harmful, humans do need some sunlight for the synthesis of vitamin D which takes place in the skin. Dark-skinned people living in regions of low sunlight today are sometimes known to suffer from a deficiency of this vitamin, and it is reasonable to assume that, as *Homo erectus* migrated northwards, a lighter skin colour evolved as an adaptation to the weaker sunlight.

Two points need to be stressed about the move into Europe and Asia. Firstly, *Homo erectus* was not necessarily driven by a strong migratory urge to travel to new unoccupied lands. They were nomadic people, and an unconscious drift of a mere 20 kilometres (12 miles) per generation would have been sufficient to cover, for instance, the 14,000 kilometres (8,700 miles) between Nairobi and Peking within 20,000 years. On the time-scale with which one is dealing, this is a relatively rapid event. Secondly, although the colonization of new continents is significant in that it expresses the new capabilities of *Homo erectus*, the majority of our ancestors probably remained in Africa. Until about 100,000 years ago, there were, perhaps, ten times as many people in Africa as there were in Eurasia, and during the periodic Ice Ages of the last million years, that figure may have been nearer to twenty times as many.

Improved tools and organized hunting

The principal cultural marker of *Homo erectus* is the Acheulean tool technology, named after St Acheul in France where it was first recognized. The Acheulean toolkit is a simple but versatile range of chopping, cutting, piercing and pounding tools, which were presumably used for butchering meat and preparing plant foods. The characteristic artifacts of the toolkit in Africa are the teardrop-shaped handaxe and the long-edged cleaver, implements that are extremely useful in butchering carcases, but may well have been employed on plant material too.

One of the startling aspects of the technology of *Homo erectus* was its lack of change over long periods of time. Elements of the toolkit emerged first about one-and-a-half million years ago, and the basic design persisted until about 200,000 years ago in Africa when it was replaced by the more complex Middle Stone Age technology. In western Europe the Acheulean continued even longer, remaining the major design feature of stone tools until as late as 100,000 years ago, when *Homo erectus* had begun to give way to early *Homo sapiens*.

Throughout the million-year span of the Acheulean technology there was no marked refinement to be seen. Indeed, some of the later examples of the technology appear simple and crude compared with some earlier material. Toolkits did differ from region to region, perhaps reflecting local traditions, but the variation was not dramatic. For instance, the difference between the Acheulean of East Africa and that of France half-a-million years ago was less than the stylistic variation that occurred between two adjacent European river valleys 10,000 years ago. Such a restricted degree of design expression over both time and space may imply limited skill and imagination in *Homo erectus* compared with *Homo sapiens*.

Although there are not enough specimens from which to draw firm conclusions, it does appear that there was a small but significant increase in brain capacity in the later populations of *Homo erectus*. The rise was from something close to 900 cubic centimetres (54 cubic inches) in early *Homo erectus* to around 1,100 cubic centimetres (66 cubic inches). This growth in brain size may have followed from an increase in the size of the body, but it may also have carried with it enhanced brain power. One hint of greater mental capacity comes from evidence of specialized hunting towards the end of the *Homo erectus* period. A site at Olorgesailie in Kenya and two sites in Spain, Torralba and Ambrona, provide the evidence for this specialized hunting.

Olorgesailie is a remarkable prehistoric site, lying in the floor of the Rift Valley some 50 kilometres (30 miles) southwest of Nairobi. Today the area is covered with grass and thorn bush and it is hot and dry, but half-a-million years ago a lake sparkled in the sun, attracting game from many parts of the valley. The lake sediments preserved evidence of an unusual hunting activity, and, mainly through the work of Glynn and Barbara Isaac, a total of over sixty extinct giant baboons have been unearthed from a spot approximately 20 metres by 12 metres (65 feet by 40 feet). More than 10,000 beautifully shaped handaxes were littered among the remains, thus dispelling any notion that the animals had died from natural causes. The collection undoubtedly speaks of an extraordinary hunt or, more likely, a series of hunts, carried out by our ancestors some half-million years ago.

The hominids of Olorgesailie must have concentrated their hunting efforts on these giant baboons for some good reason. The baboons cannot have been easy prey as they were equipped with large canine teeth and were powerful animals, the size of a female gorilla. A skilled co-operative hunt, possibly under cover of night, would have been necessary for repeated success. Pat Shipman of Johns Hopkins University, Baltimore, has detected a curious and consistent pattern in the butchering of these animals, as revealed by cut-marks on the bones. What is odd is that the animals were not butchered in the most efficient way. This, she suggests, may indicate that the regular hunting of these giant baboons was an important part of some kind of ritual. This may or may not be the case, but the stones and bones of Olorgesailie indicate an ability for specialized hunting that is not obvious earlier in the archaeological record.

Signs of organized, large-scale hunting also come from two somewhat younger sites in the Sierra de Guadarrama in central Spain. The hills of Torralba and Ambrona stand on either side of a small stream, which forms part of the only pass for many miles through the mountains. There were once swamps in the valley floor, and there, over 300,000 years ago, *Homo erectus* butchered the carcases of many large animals that had become mired in the swamp. At Torralba, for instance, the remains of at least thirty elephants, six rhinos, twenty-five red deer, twenty-five horses and ten wild oxen have been found. The tally at Ambrona is similar. The interesting point is that there are indications that *Homo erectus* did not simply wait for these animals to blunder into the swamp, but took a more active part in the process, and drove the animals to their death.

The evidence of the two sites points to the co-ordinated actions of groups of intelligent people. And there are signs of an interest in the animals beyond that of basic subsistence. At Torralba, for instance, the almost complete left side of an elephant skeleton was arranged as if for an exhibit. And at Ambrona two lines of elephant bones are set perpendicular to one another, forming a deliberate 'T' pattern. Yet another mystery is the fate of the elephant skulls. Of the many animals that were butchered at the sites, the skull of only one of them remains, and this is associated with the T-shaped pattern at Ambrona. This all suggests that a certain element of

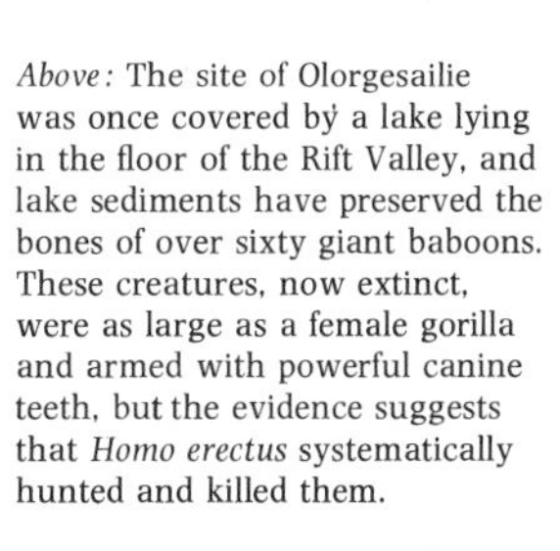

Above: The site of Olorgesailie was once covered by a lake lying in the floor of the Rift Valley, and lake sediments have preserved the bones of over sixty giant baboons. These creatures, now extinct, were as large as a female gorilla and armed with powerful canine teeth, but the evidence suggests that *Homo erectus* systematically hunted and killed them.

Right: Over 10,000 carefully made handaxes were found at Olorgesailie, intermingled with the bones of the giant baboons. This photograph shows two limb bones and three of the teardrop-shaped handaxes that were the principal component of the *Homo erectus* toolkit.

ritual was emerging in association with the large-scale hunting activities of *Homo erectus*.

Peking Man

Hints of archaic ritual also appear in one of the most famous *Homo erectus* sites in the world: the Peking Man cave in northern China. An hour's journey by road southwest from Peking is the small industrial town of Choukoutien, and nearby, two limestone hills overlook a river. From the town one can see that the hill to the east, known as Dragon Bone Hill, is deeply scarred: this is the result of early mining for limestone, followed by the removal of half-a-million tonnes of archaeological deposits from an ancient cave. The remains of perhaps forty *Homo erectus* individuals have been retrieved from those deposits, in the form of crania, fragments of skulls and jaws, and parts of limb bones. Twenty thousand stone tools made by the cave dwellers some half-million years ago have also been recovered.

The story begins in the early 1920s with Davidson Black, head of the department of anatomy at the Peking Union Medical College. Davidson Black was firmly convinced that primitive man would one day be found in China, and he was equally determined that he would be the discoverer. The Chinese had an interest in fossil bones too, but for medicinal rather than scientific purposes. They ground fossil bones to a powder and mixed this with other ingredients. Fossils, or 'dragon bones' as they were called, were thought to have strong curative powers. Indeed, it was the local knowledge of Dragon Bone Hill's rich store of fossils that first attracted Davidson Black to the Choukoutien site.

The hominid finds from the cave were meagre at first, but this did not dampen Davidson Black's enthusiasm. With just two isolated teeth to back his claim, he announced to the scientific world in 1927 that a new species of primitive man had been discovered: *Sinanthropus pekinensis*, 'the Chinese man of Peking'. Fortunately more fossils turned up, and, on 2 December 1929, Pei Wen-chung, head of the Institute of Vertebrate Palaeontology and Palaeoanthropology in Peking, discovered the first skullcap.

A worker involved in the project described the events in the following way: 'It all took place after 4 o'clock in the afternoon. We had got down about 30 metres [100 feet] deep and only three men could stand at the bottom of the hole. It was there the skullcap was sighted, half of it embedded in loose earth, the other half in hard clay. The sun was almost set and the light getting poorer. The team debated whether to take it out right away or to wait until the next day when they could see better. The agonizing suspense of a whole night was felt to be too much to bear, so they decided to go on. It was well done. A small piece of the bone cracked in the course of the digging, but no serious damage was done, and the restoration was easy. The whole piece had to be dried to make it firm enough to take to the research unit in Peking. The best method available was drying over a charcoal fire, and that was how it was done. This process took a day and two whole nights.' That skullcap was the first of half-a-dozen similar specimens that were eventually to be recovered from the cave deposits.

The hillside cavern had originally been massive, measuring 140 metres (460 feet) from east to west, 40 metres (130 feet) north to south, and 40 metres (130 feet) from floor to ceiling. As with the limestone caves of the Transvaal in South Africa, the Choukoutien cave provides discrete glimpses of the past because it was open only periodically to the outside world. Thirteen such periods are to be seen in the separate layers in the cave, giving glimpses of life half-a-million years ago in the lowest layers, and

Chinese archaeologists drying out the first skullcap of Peking Man, a few hours after it was found in 1929. The fragile cranium was held over a charcoal burner for thirty-six hours to make it strong enough to survive the journey to Peking.

rising to 200,000 years ago in the higher deposits. Remains of Peking Man are found in many layers of this sequence. In the very top of the cave known as the Upper Cave, there are skeletons of modern humans that were buried just 15,000 years ago.

During the many thousands of years that *Homo erectus* lived in the environs of Choukoutien, the climate fluctuated between being rather similar to today, that is, very hot in summer and very cold in winter, and being somewhat warmer. The environment offered a mosaic of forest-covered mountains, open plains, rivers and lakes. The forests were a mixture of pines, cedars, elms, hackberries and Chinese redbuds and were inhabited by bison, sabre-tooth tigers, leopards, brown bears, black bears, red dogs and wolves. On the plains to the southeast a lush growth of goose-foot, lily, sagebrush and pyrola sheltered populations of horse, sika deer, elephant, woolly rhinoceros, striped hyaena and cheetah. This was a rich, if occasionally rather chilly, environment in which to pursue a hunting-and-gathering economy.

One of the most important discoveries in the Peking Man cave was the first unequivocal evidence of the use of fire by hominids. This appears even in the earliest times of occupation. Thick layers of ash, burned and charred bone, all restricted to what were unmistakably hearths, were found in several parts of the cave. Peking Man quite clearly retreated to the shelter of the cave during the cold winter months, keeping himself warm and other animals at bay by the flames of a fire. The cavern was certainly big enough for a band of twenty or so individuals, and they appear to have returned to the cave year after year. One of the ash layers is 6 metres (20 feet) thick which indicates a very long sequence of seasonal occupations. The inhabitants of the cave may also have begun the habit of cooking their meat before eating it, a development that would certainly have eased the business of chewing their food.

Curiously, all the hominid crania that have been found in the Choukoutien cave appear to have suffered the same kind of damage: the underneath surface of the brain case, through which the spinal cord passes, is missing.

Woo Ju-Kang, head of the Institute of Palaeontology in Peking, at the cave near Choukoutien where Peking Man once lived. His excavations have unearthed new specimens of *Homo erectus* from the cave.

Some people have interpreted this as indicating that Peking Man was a cannibal who ate the brains of his brothers. If indeed Peking Man did occasionally consume human brains it clearly was not in ravenous hunger, for a much easier way of reaching the brain tissue is simply to smash the skull open. The damage on the Choukoutien skullcaps gives the impression of careful and deliberate entrance to the cranial cavity, perhaps an act of ritual.

Did small groups of *Homo erectus* half-a-million years ago sit around a glowing fire, reverentially eating the brain of a loved-one so as to ensure a continuation of his or her spirit? Woo Ju-Kang, who now heads the Institute of Palaeontology in Peking, is cautious on the matter: 'It is of course possible that the skulls were opened for some form of ritual cannibalism. But it may also be that the skulls were damaged as they became buried and preserved in the cave deposits. The bottom part of the cranium is delicate and breaks easily. There is no certain way of answering this question, but I consider that accidental damage is the more likely.'

Sadly, Davidson Black did not live long enough to gaze upon the rich store of fossils that were excavated from the cave. But his death in March 1933 also spared him from knowing of the catastrophic disappearance of the entire Peking Man collection.

The German anatomist Franz Weidenreich took over from Davidson Black, and between 1935 and 1937 some remarkable discoveries were made under his supervision. Work at the site had to cease in 1937 because of hostilities with Japan, and for the next three years Franz Weidenreich spent much of his time studying and, as luck would have it, making excellent casts of the fossils. It was believed that, in the event of war with Japan, the invaluable Peking Man fossils would be at risk. Plans were therefore made to ship the fossils to safety, a task entrusted to the American Marines. Unfortunately, somewhere on a train journey between Peking and the port of Chinwangtao, where the steamship *President Harrison* was waiting for them, the boxes containing the fossils disappeared. Despite strenuous efforts by many people, the original fossils of Peking Man have never been seen again.

With the shadow of the lost fossils ever-present Woo Ju-Kang and his colleagues reopened the excavation in 1949, only to be interrupted seventeen years later by the Cultural Revolution. That period of intermittent work yielded a number of *Homo erectus* teeth, most of a lower jaw, and two parts of a skullcap that fitted together with casts of several fragments unearthed thirty-three years previously. These fragmentary pieces are all the fossil material that exists of Peking man, with the exception of two teeth from the original excavation which had been sent to Sweden and are still there.

The search is, however, once again underway at Choukoutien. Woo Ju-Kang organized an international conference in Peking at the end of 1977 to mark the fifty years since excavations had begun at the famous site. In May the following year the excavators once more moved in. 'It is a multidisciplinary project,' he explains. 'There are seventeen organizations involved in the research and our aim is to build up a record of the environment and climate at the time that Peking Man lived in the cave. Of course, we would also like to find some more good specimens of Peking Man himself.'

The loss of the first Peking Man fossils was a tragedy for prehistorians. By good fortune, excellent casts had been made of them, and a cast of one skullcap is shown here, behind reconstructed fragments of a skullcap from more recent excavations. Both skulls show clearly the prominent brow-ridges typical of *Homo erectus*.

A Mediterranean camp site

One important *Homo erectus* site, which may be of about the same age as the Choukoutien cave, is Terra Amata, which overlooks the Mediterranean near to Nice's commercial harbour. Uncovered initially by bulldozers, the site is now entombed in the foundations of a block of luxury apartments. Fortunately, a five-month rescue operation led by Henri de Lumley, then of the University of Aix-Marseilles, retrieved a massive amount of archaeological information that tells the story of life beside the Mediterranean some 400,000 years ago.

The climate was chillier at the time of Terra Amata, and the Mediterranean was 25 metres (80 feet) higher than it is today. Fir trees and Norway pines on the nearby Alps grew further down the slopes than they do now, and heather, sea pine, Aleppo pine and holm oak covered Mont Boron and its neighbouring coastal peaks. A small cove had been carved in the limestone of Mont Boron's western slope. Within the cove a sandy, pebble-strewn beach offered shelter from the winds. A nearby spring bubbled a constant stream of clear, fresh water. An attractive location to be sure, and Henri's excavations reveal that at least one hunter-gatherer band chose it as a brief springtime home for a number of years.

The most interesting results of the excavation were the traces in the sand of a series of eleven large, carefully constructed dwellings, each built on roughly the same spot as the previous year's. They were oval in shape and measured about 12 metres (40 feet) long by 6 metres (20 feet) wide. They were constructed from walls of young branches supported in the centre by a row of sturdy posts. The people of Terra Amata placed large stones around the base of the walls so as to add extra support against the northwest wind.

The importance of the discovery is not so much the construction itself, but the glimpses of activity within. A hearth was built near the centre of each hut. A scatter of stone flakes indicated the work of a tool-maker, and an area in the middle which was clear of flakes showed the place where he squatted as he worked. One set of eleven flakes could be reassembled to form the original pebble: clearly the tool-maker knapped some flakes and then made little use of the products. The hut dwellers used animal skins for comfort, probably both for sitting on and for sleeping on. A curious depression in the sand could have been made by a long-vanished wooden bowl. Most intriguing of all, though, are traces of worn ochre of the sort that French historian François Bordes has suggested was used for body-painting.

Remains of red deer, elephant, an extinct species of rhinoceros, mountain goat and wild boar reveal the hut dwellers' taste for meat. Many of the animals brought to the camp were young, indicating that they were hunted rather than scavenged. Shells of oysters, mussels and limpets show that these people made some use of the resources of the sea. As usual, there is little to suggest what plant foods were collected and eaten although the groves of bushes and trees beyond the beach must surely have provided many items of food.

Henri knows that the seaside camp was occupied during late spring from an analysis of pollen contained in the occupants' fossilized faeces. The faeces contained the pollen of broom which flowers only in late spring. He also knows that each year the newly built camp was inhabited for just a short while: a longer occupation would have compacted the sand in the area of the hut. The palisade walls fell quickly after the group left camp, although on one occasion at least, they appear to have burned it down. 'In the fall the winds covered the living floors, the levelled palisades and the rest of the camp debris with a layer of sand perhaps two inches deep,' Henri explains. 'The rains then spread out the sand and packed it down, so that when the hunters returned to the cove the following year the evidence of their earlier stay had been almost obliterated.'

The people who made Terra Amata their brief springtime home over a

A reconstruction of the Terra Amata hut. The hominid band who made Terra Amata their springtime home year after year built a hut of sturdy posts, supported by large stones for protection against the wind. Over the posts were laid young branches, forming the walls. Each year, after the hut was abandoned it was covered by the drifting sands which preserved the evidence of the hominids' visit.

period of years were clearly following a well-established way of life as nomadic hunter-gatherers. Their knowledge of the living world as a potential food resource was extensive and their skill in exploiting that resource was highly developed. Although their stone-tool technology was not particularly sophisticated in physical form, they were probably very skilled in its use. Acts of ritual, perhaps associated with the changing seasons, perhaps with the changing phases of life, may have been an important part of that ancient society.

One element which I have not so far mentioned, which leaves no trace in the archaeological record, but which may have been crucial to the operation of this highly complex lifestyle, is language. It seems very likely that *Homo erectus* had command of some form of spoken language, however rudimentary. The evidence for this is examined more closely in the next chapter.

8 The Birth of Language

There is a cartoon that shows two prehistoric men wondering, now that they had learned to talk, what exactly they would talk about. Strange as it may seem, modern psychologists and prehistorians are no wiser than the cartoonist: they simply cannot agree on why language evolved in the first place. 'Speech is the best show man puts on,' proclaimed linguist Benjamin Lee Whorf, while eminent biologist, George Gaylord Simpson, called it 'the single most diagnostic trait of mankind'. But no one dares to be so dogmatic on the subject of what evolutionary pressures produced speech.

Humans in their modern form, *Homo sapiens sapiens,* arose at least 40,000 years ago, and it is reasonable to assume that a refined and precise form of language was on their lips. The question is, when did language first begin to evolve in our hominid ancestors? In an attempt to answer this question several indirect lines of enquiry have been followed. Firstly, the potential for speech in our closest relatives, the chimpanzees, and in the less closely related gorillas, has been investigated. Secondly, the faint images of brain structure that are seen on the interior surface of fossil skulls have been studied. Thirdly, stone tools have been looked at more closely for clues about human speech. Fourthly, the paintings, engravings and carvings of the peoples of the last Ice Age may offer an insight into mental processes and the degree of development of language at that time.

The talking apes

Apes, of course, cannot talk, but they can, it seems, understand and use sign language. The idea of teaching some form of sophisticated communication to non-human primates is not new. Samuel Pepys made the following entry in his diary for August 1661 after seeing a baboon (or it might have been a chimpanzee) in London: 'I do believe it already understands much English; and I am of the mind that it might be taught to speak or make signs.' Almost a century later a French philosopher suggested that apes might be able to master a manual language. The idea was proposed at intervals over the years, and in 1925 the great American primatologist Robert Yerkes, outlined the notion specifically: 'Perhaps [chimps] can be taught to use their fingers, somewhat as does the deaf-and-dumb person, and thus acquire a simple nonvocal sign language.' But forty years went by before anyone put the scheme to the test.

In 1965, Allen and Beatrice Gardner were lecturing in psychology at the University of Nevada. They saw a film of a chimpanzee, Viki, who had been taught to say what sounded approximately like 'Mama', 'Papa', 'cup' and 'up', the meagre results of four years' training by Keith and Cathy Hayes. They were interested in the problem of language in non-human primates, but decided that teaching a sign language might be more fruitful. They

In an exciting new approach to the study of language scientists have been teaching our nearest relative, the chimpanzee, to use sign language. Some impressive results have been obtained by Roger Fouts and others working at the University of Oklahoma. Seen here is a chimpanzee called Moja, making a sign for food.

took charge of an eleven-month-old chimpanzee called Washoe and found that she could indeed pick up the specific 'words' of American Sign Language, Ameslan. The progress was slow but steady. Her vocabulary gradually increased and she appeared to enjoy her new skill as a way of communicating under social, as opposed to experimental, surroundings.

Roger Fouts, who is currently Washoe's guardian at the University of Oklahoma, was working with the Gardners at Nevada in the early days of the project. 'We worked on the premise that language is something that occurs in families,' he explains. 'Instead of forcing Washoe through regimented teaching and testing routines, we adopted the natural approach: language is something you *learn*, not something you're taught. So Washoe was immersed in an environment of sign language, just as a child is constantly exposed to verbal language. As scientists, we sometimes felt it was rather odd because we seemed to spend much of our time babysitting with Washoe: making her breakfast, giving her baths, playing hide-and-seek with her and so on. I had questions about the project at the time, but I now recognize that it was a crucial move. Language, you see, is a *social* behaviour; it's about relationships with other individuals.'

Moja, like several other chimps, has now mastered sign language. With her is Roger Fouts, who also worked with the first 'talking chimp', Washoe. Believing that language is most readily learned in the social context of a family, he and other scientists engaged in the project spend a lot of time feeding, bathing and playing with each chimp while trying to teach it to make signs.

Washoe is no longer the only ape to have mastered sign language. There are now a number of chimps who have gained quite large vocabularies, and in the course of filming I was able to 'talk' with one of Washoe's companions, called Moja. Several primate research laboratories throughout America house chimps who communicate with their caretakers via various forms of formalized symbolic languages. For instance, at Princeton University a chimp called Sarah converses with researcher David Premack using a language composed of plastic shapes. At the Yerkes Regional Primate Center in Atlanta, Duane Rumbaugh has taught another chimp, Lana, to manipulate an artificial computer-based language called Yerkish. More recently, there have been experiments in teaching sign language to gorillas and these have also met with some success.

Language is, of course, constructed from words and from the rules of grammar which determine their order. Words are arbitrary inventions of the human mind. A tree, for example, is labelled in English by the word 'tree' only because we agree that it is, not because the word looks or sounds anything like the object. Once some form of arbitrary label has been fixed to an object, one has a very useful system: in this case the label 'tree' can be used on anything from an enormous redwood to the most diminutive bonsai. Each variety of tree can have its qualifying name of course, but they are brought together as an identifiable group by the term 'tree'.

The recognition of an object or an action as being one of a group is a mark of a certain level of intellectual ability. It requires efficient memory combined with analysis and flexibility. A chair looks different from different angles, but our brains classify all the various images as 'chair', not as a series of different objects. Indeed, the images are recognized as different views of the same chair, not a hundred different chairs. I may seem to be labouring the point, but it is important to emphasize that unless a brain can operate such a classification system, it will not be able to use the symbols that are the currency of language.

The fact that chimps and gorillas *can* name objects, and generalize those names to other, similar objects, implies that they may have the intellectual capacity to use language. But what of sentences? Chimpanzees tend to produce sequences of words in which each individual item is valid but which together make nonsense, because they do not conform to the rules of grammar. Many linguists, particularly Noam Chomsky, see basic rules running through all human languages. Different peoples may speak different languages through cultural separation, but underneath the words lies a similar grammatical structure. The reason, they suggest, is that language is the product of certain specific brain structures or functions. Noam Chomsky and his followers see language as an exclusively human trait.

Certainly, apes that have learned sign languages show little grasp of sentence structure. Washoe is as likely to sign 'drink gimme' as 'gimme drink'. Does this mean that chimps and gorillas are simply bright enough to learn the naming game but possess no real facility for language as Noam Chomsky would define it? On the face of it this conclusion seems inescapable. 'But,' retorts Roger Fouts, 'it depends on how you view language. Writing, for instance, is a highly idealized form of language: it's highly structured and therefore rather artificial. I see language as a continuum, stretching from the written word, which is what Chomsky recognizes as language, to the utterance of a single word between two people who know each other so well that it carries the import of a paragraph. Gestures, expressions, tone of voice, they're all part of language. Eighty per cent of the meaning in a face-to-face conversation passes non-verbally.'

There is in fact an increasingly bitter debate among psychologists concerning the so-called talking chimps, with the critics claiming that the apes have demonstrated only what was known already: that they are intelligent animals. One researcher, Herbert Terrace, who has conducted a long-term sign language project with a chimp called Nim Chimsky (reference to Terrace's mentor), has now reluctantly concluded that the apes have no language ability whatsoever and that all the projects of the past decade tell us nothing about human language, except that no animal has it. 'The trouble with this position,' says Roger Fouts, 'is that it applies a rubber ruler. The chimps are dismissed as having no language because their signing doesn't follow the strict rules of grammar, and yet when a child says "ball mine" the phrase is interpreted as "please may I have my ball". You can't have one standard for one situation and another for the second.'

The human brain is divided vertically down the middle into two hemispheres. There's a difference in the way the two hemispheres 'think', and this is crucial to the way that Roger Fouts views the question of language. The left hemisphere tackles the rigid analytical problems of the world: it is the side with which you would solve a mathematical problem. The type of mental processing in this hemisphere is described as being 'sequential'. By contrast, the right side of the brain takes things as a whole: it is the part of your brain with which you appreciate a beautiful painting or come up with a brilliant flash of inspiration. The mental processing here is termed 'simultaneous'. Roger Fouts exemplifies the differences in these words: 'Einstein got $E = mc^2$ by simultaneous processing in his right hemisphere, and then spent the rest of his life trying to analyse its implications using sequential processing in the left hemisphere.'

The Western world is very much a left-hemisphere dominated environment: its rules, structures and teaching are primarily governed by a logical, analytical approach. It is a world in which logic is revered and intuition is not trusted. 'Our whole experience is immersed in the world of sequential processing,' says Roger Fouts, 'and foremost in this is the way we structure language. It is seen as a logical, structured system. There's a good deal of evidence to show that the Chinese and Japanese, who use a pictographic language, are much less sequentially biased than we are. It's no accident they are much more sympathetic to the language of the body and the power of intuition than we are.'

Throughout human evolution, Rober Fouts argues, there has been a steady trend towards more ordered ways of interacting with the world. 'Tool-making, speech, social and economic organization, all require a more sequential mode of mental activity than happens in other animals. It's not surprising that that part of our brain predominates. But it's a question of experience as much as genetic endowment. The person brought up in an environment in which the analytical approach is emphasized will view the world through sequential lenses, so to speak. It's interesting that Washoe, who has been immersed in a human world for most of her life and has been exposed to the business of learning a formal language, now tackles problems more like an adult human than like a chimp. Her experience is that of a logical, analytical world, and that's how her mental processes have been shaped.'

The work with ape sign language is therefore serving to sharpen our ideas about the nature of language. It is clear that these apes, though unable to form structured sentences in sign language, do have in their heads the basic seeds of language. It is not unreasonable to suppose that evolutionary processes have, over a long period of time, developed man's complex verbal abilities from just such simple seeds.

The imprint of the brain on fossil skulls

Although apes' brains are not as different from those of humans as has frequently been implied, there is some difference, both in size and shape. One of the prominent features of human anatomy is the globular skull that houses a brain of considerable size. What is important about brain size is its ratio to body size, and on this scale humans show up as being far better endowed than any of the apes.

As the fossils from Africa illustrate, the hominids of about three million years ago had brains that were very ape-like in size. Presumably, the hominids were living very differently from their ape relatives because they were inhabiting more open environments and they were also walking around on two legs. But whatever their lifestyle, it does not seem to have demanded a significantly expanded brain. Not until two million years ago is there firm evidence of increasing brain power, with the emergence of *Homo habilis,* whose cranial capacity was close to 800 cubic centimetres (49 cubic inches). This creature had a brain that was almost twice as big as a chimp's but with a very similar body size. With the evolution of *Homo erectus* around one-and-a-half million years ago there was a further boost to brain power, the cranial capacity increasing over a period of a million years to more than 1,000 cubic centimetres (61 cubic inches). The modern human capacity of 1,360 cubic centimetres (83 cubic inches), on average, was reached within the last 100,000 years.

A word of warning about the implication of brain size in humans: the variation is enormous and is not necessarily related to intelligence. Jonathan Swift, for example, had a brain of around 2,000 cubic centimetres (122 cubic inches), while Anatole France managed more than adequately with a mere 1,000 cubic centimetres (61 cubic inches). Clearly, there is more to intelligence than the size of the brain, but exactly what it is remains a mystery.

On the face of it, the human brain seems to have developed relatively late on the evolutionary scale. If *Homo* first emerged over four million years ago, there must have been a long period of time during which its brain remained substantially primitive. Or did it? Ralph Holloway, an anthropologist at Columbia University, New York, has been studying the brains of our ancient ancestors for more than a decade, and has come to some rather remarkable conclusions. Ralph obviously cannot study the brains of the hominids themselves since brains do not fossilize. But what he does have is the inside surface of the fossilized cranium. Remarkable as it may seem, the brain does leave its signature on the inner surface of the cranium. For the most part the details are extremely faint, but there is enough to be able to make out the general organization.

Each half of the brain is divided into four parts, or lobes, and one can make some general statements about the function of each lobe, though these are, necessarily, rather simplistic. The one at the front, the frontal lobe, is responsible for controlling movement and some aspects of the emotions. The one at the back, the occipital lobe, handles vision and the area at the side, the temporal lobe, is important for memory. Above the temporal lobe is the parietal lobe which has the crucial role of comparing and integrating information that flows into the brain through the sensory channels of vision, hearing, smell and touch. Roughly speaking, a 'human' brain is one in which the parietal and temporal lobes predominate whereas in ape brains these areas are much smaller.

With these rough guidelines in mind, Ralph examined the architecture of some australopithecine and *Homo* brains and came to the following conclusion: 'The basic shape of the human brain is clearly evident in the

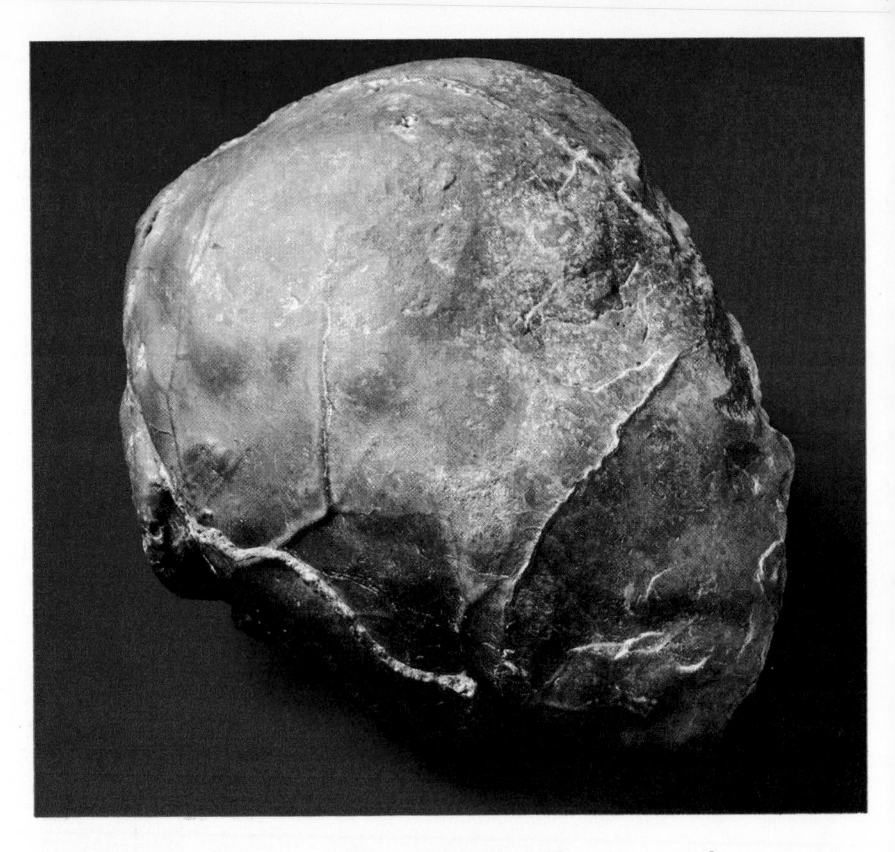

An endocast of the skull of a robust australopithecine from the Swartkrans cave. Such endocasts can reveal much about the shape of the brain, and the size of the swelling known as Broca's area which plays a vital part in speech.

hominids of *at least* two million years ago.' This was a rather surprising discovery, as the size of the australopithecine brain was not dramatically different from that of a chimp or gorilla brain. In preliminary work on a fossil from Hadar, Ralph finds the same human brain pattern emerging in spite of the small size of the brain, and this in a hominid that lived more than three million years ago. But what of language?

In modern man, a region towards the front of the left hemisphere of the brain, known as Broca's area, co-ordinates the muscles of the mouth, tongue and throat when we speak, and a second region at the side of the left hemisphere, Wernicke's area, is responsible for the structure and sense of our language. Wernicke's area receives information from the ears and the eyes, and it is located close to a major 'association area' of the cerebral cortex which integrates and compares the incoming information from all the senses. Sentences that issue from our lips have been organized according to grammatical form by the neural programmes in Wernicke's area, but the actual muscular movements necessary to produce the sounds are controlled by Broca's area. The anatomical effect of this is that the left hemisphere is rather larger than the right, and there is a detectable lump over the region that houses Broca's area. (In some left-handed people, the vital areas are switched to the right side of the brain.)

Can we trace the origins of language in the human ancestral line by looking for the imprint of Broca's area in ancient crania? Unfortunately, the quest is not so simple. Firstly, as Ralph points out, 'the fossils seem to be plain conspiratorial: during fossilization they often become distorted at one side or another so as to make the task of comparing the two hemispheres virtually impossible.' And secondly, apes' brains are usually bigger in one hemisphere than the other (though it is not predominantly the left side that is enlarged), and there is a swelling where Broca's area should be. What this means is something of a mystery. The swelling is not as pronounced as in humans, but its presence is tantalizing.

Ralph has looked at a number of fossil skulls and in skull 1470, an early representative of *Homo habilis,* there is the impression of a distinct Broca's area, larger than that in the apes, but not as definite as that of modern humans. The impression of Broca's area is more strongly marked in *Homo erectus* than it is in *Homo habilis.* Does this mean that our ancestors of two million years ago had rudimentary language? It is difficult to be sure. Furthermore, the australopithecines also display an enlargement in this part of the brain. Were they, too, beginning to use language to communicate with each other?

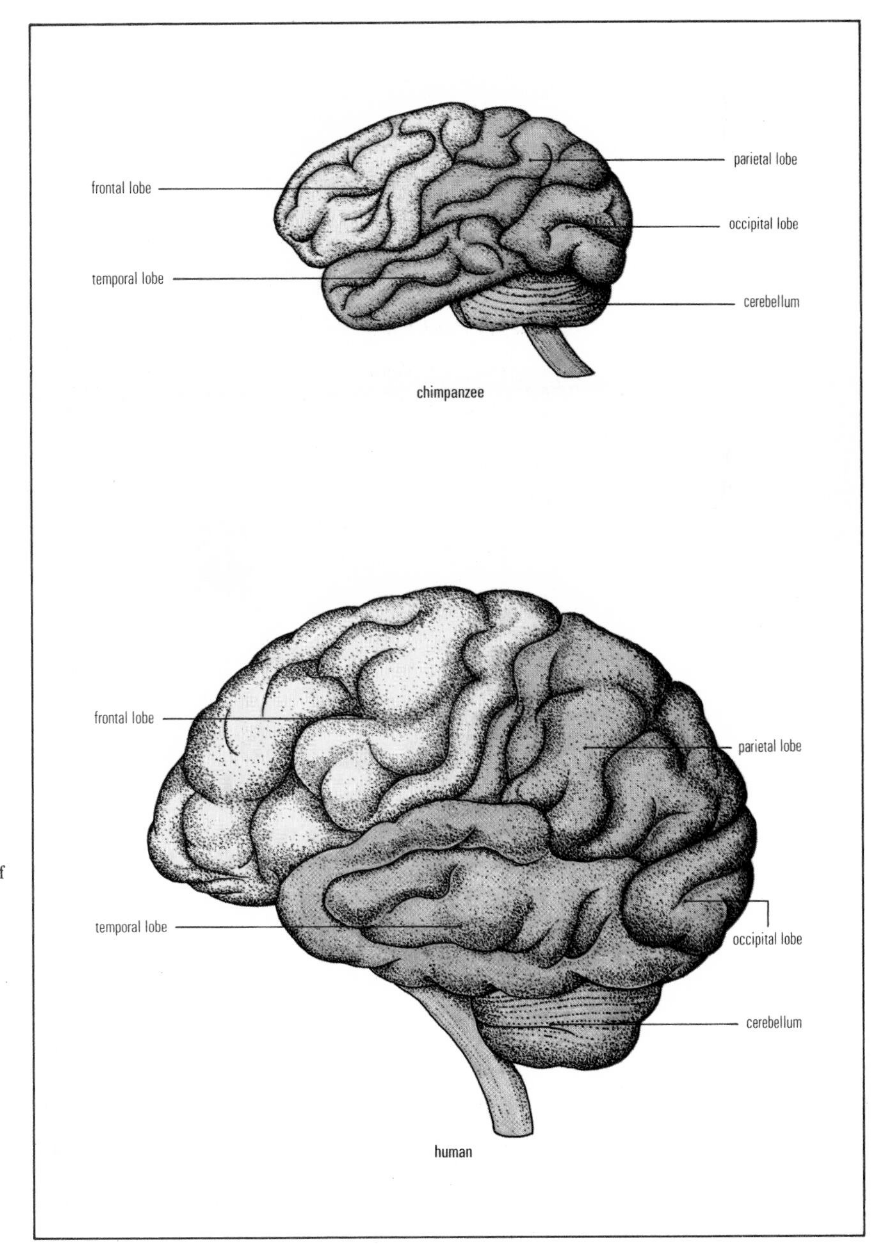

The australopithecines' brains were not much larger than those of apes, but with the emergence of the *Homo* line there was a steady increase in the size of the brain. Size, however, is not the only significant factor: the shape of a human brain differs substantially from that of a chimpanzee, with certain brain lobes becoming increasingly dominant. These are the temporal lobe, responsible for memory, and the parietal lobe which integrates information received from the senses. This change in shape is evident at a very early stage in hominid evolution, and, surprisingly, even the small-brained australopithecines had a basically human brain form.

The link between tools and language

Speech is, archaeologically, an invisible aspect of human behaviour. Stone tools, by contrast, survive in the record supremely well. Can their shape, form and association give an insight into past language? There may in fact be important messages in those ancient implements.

Some years ago Glynn Isaac addressed an international conference on the origins of language in the following terms: 'Asking an archaeologist to discuss language is rather like asking a mole to discuss life in the treetops. The earthy materials with which archaeologists deal contain no traces of the phenomenon that figure so largely in a technical consideration of the nature of language. . . . If, however, all the trees have been cut down and all that remain are the roots, then the mole may not be such an inappropriate consultant. So it is with the history of language development.' Glynn suggests two approaches to the problem: 'The first involves the scrutiny of the record of developing protohuman material culture remains. . . . Stone artifacts are the best and most persistent long-term markers, but during the last five per cent of the timespan we can also deal with more fancy evidence such as burials, ornaments, art, notations, cult objects, structures and so forth; the second approach involves taking evidence of economic behaviour as a whole.'

Left: Crude stone tools are found in the fossil record from about two-and-a-half million years onwards. They lack the standardized form of later tools but they have been shown to be highly efficient when used for butchering carcases.

Right: A teardrop-shaped handaxe of the Acheulean industry, which emerged about one-and-a-half million years ago and is associated with *Homo erectus*. These handaxes have a symmetrical form not seen in the haphazardly knapped stone tools of earlier times.

The archaeological record of tools begins between two-and-a-half million and two million years ago. Crude tools litter the record until around one-and-a-half million years ago, when the teardrop-shaped handaxes of the Acheulean industry appear. These introduce an element of

symmetry and a sense of purpose that previously was absent. The Acheulean technology continued until 200,000 to 100,000 years ago, when a new and more economical method of striking flakes, the Levallois technique, was invented. From this time until about 40,000 years ago the Mousterian industries, which relied upon the Levallois technique, flourished and they were accompanied in a social context by the first signs of burials, grave offerings and cult objects. Regional differentiation of form, which had begun to become apparent in earlier times, now became strongly pronounced. From 40,000 years onwards variations in stone-tool cultures and evidences of artistic expression increased greatly. The progression through this huge tract of time was at first very slow, but steadily gathered pace.

If one considers the stone implements made by our ancestors over the past couple of million years, an interesting paradox emerges: although there is a steady increase in the number of identifiable tool types, together with a refinement of the individual implements, the *range* of artifacts does not broaden significantly until about 40,000 years ago when new and more delicate types of tools appear. In other words, even in the very earliest toolkits one can find edges, points, surfaces and so on that are features of the basic implements of stone-tool cultures right up to the Late Stone Age.

There is a difference, however, in that as time moved on the stone-knappers clearly had in their heads a repertoire of distinctive target shapes which, through manual skill and keen judgment, they could regularly achieve. Stone-tool manufacture became very much a matter of imposing on the stones a form which had been thought of in advance, although the highly differentiated, standardized tool forms had no great advantages over the more haphazardly knapped, crude tools of earlier periods. The basic details of the standard form were adhered to although they were not critical to the usefulness of the tool.

Glynn Isaac proposes that with each step in the cultural evolution of *Homo sapiens* individuals experienced a greater sense of order: 'Probably more and more of all behaviour, often but not always including tool-making behaviour, involved complex rule systems. In the realm of communication, this presumably consisted of more elaborate syntax and extended vocabulary; in the realm of social relations, perhaps increasing

Tools of the Mousterian industry: (left) a stone core from which flakes were struck, (centre) a side-scraper used for cleaning hides, (right) a point, a tool assumed to have been used as the tip of a spear or arrow. The appearance of these tools about 100,000 years ago was accompanied by the first signs of ritual burials, grave offerings and cult objects. At the same time, different styles developed in different regions, suggesting that local cultural traditions were emerging.

A flint spear-head of the 'laurel leaf' type, dating from about 18,000 years ago. This particular cultural style was restricted to a small area of western Europe, but it typifies the precise imposition of standardized form seen in later stone-tool cultures. Such form was imposed for its own sake, not because it greatly improved the usefulness of the tools.

numbers of defined categories, obligations and prescriptions; in the realm of subsistence, increasing bodies of communicable know-how.' According to this viewpoint the sophisticated tools of, say, 30,000 years ago reveal more about the mind of the individual that made them, than they do about the task for which they were intended.

This concurs very much with what Roger Fouts has proposed on the basis of his experience with Washoe: that human evolution produced a steadily more ordered world, a world in which sequential mental processes became paramount, a world in which some kind of formal language was increasingly important.

Language and art

Another product of prehistoric hands that offers an important insight into what was going on in the people's minds is art. Between 30,000 and 10,000 years ago there was intensive artistic activity, which produced a vast range of paintings, carvings and engravings.

Locked away in a bank vault in Germany is a collection of exquisitely carved statuettes dating from around 32,000 years ago. The statuettes, which include representations of horses and a crude human figure, were found in the early 1930s at a small cave site called Vogelherd in southern Germany. These objects are among the oldest forms of representational art yet known.

One of the statuettes, a 6-centimetre (2·5-inch) long horse finely carved from mammoth ivory, is as evocative an image as one can come across. But the really significant aspect of the Vogelherd horse, and of the other objects from the site, is that it was undoubtedly handled frequently over a long period of time. Its smooth edges and polished surfaces indicate that it was art that was in constant use. American researcher Alexander Marshack emphasizes very strongly the *use* of art objects as being central to their importance. And this use often involves adding new marks, such as lines, zig-zags and pits, from time to time.

One of the Vogelherd figures which might be a reindeer or possibly a horse, for the fragment is too incomplete to allow one to be certain, is marked with multiple zig-zags, arcs and many small strokes. Another of the figures represents a mammoth and this has a semicircle of crosses etched into its flank. Such patterns are common and persistent throughout Ice Age art, and may well have been of symbolic significance to the people who used the objects. Marshack describes the Vogelherd combination as an example of the 'earliest *Homo sapiens* symbolism'.

From the time of the Vogelherd figures onwards, the artistic output of our forebears in Europe and Africa was prolific. But what were its origins? There are paintings of animals on rock slabs in a cave in southern Africa dating from about 29,000 years ago. These have a good claim to be the earliest form of wall art. But engraving and carving apparently have a longer history. For instance, there is a pendant made from a reindeer's foot bone from La Quina in France which is at least 35,000 years old. A fragment of bone marked with a zig-zag motif from the Bacho Kiro site in Bulgaria stems from the same period. And from a 50,000-year-old site at Tata, Hungary, comes an intriguing object: a mammoth molar tooth that has been carved, shaped and worn smooth with use. On at least one occasion it had also been coloured red with ochre. 'Here,' says Marshack, 'the artisan planned for a "non-utilitarian" symbolic object.' The oldest engraved object so far discovered and dated takes us back an incredible 300,000 years, to the site of Pech de l'Azé in France. There, in 1969, François Bordes discovered an ox rib that had been engraved with a series of double arcs. The motif, once again, is a frequent feature of the art that was to follow more than a quarter-of-a-million years later.

As with much of archaeology, the objects available for study are those that survive through time. The Australian Aborigines, for instance, weave a

A statuette of a mammoth, one of the carved ivory figures from Vogelherd that date back to 32,000 years ago. A semicircle of crosses has been etched into the side of the animal, and similar symbols are seen on some of the other Vogelherd figures. Such marks suggest the figures had a symbolic significance, which points to the development of a sophisticated language, able to convey abstract ideas, in the people of this era.

Mysterious symbols are seen in many of the painted caves of Europe and Africa, which are between 30,000 and 10,000 years old. Such symbols are convincing evidence that modern language capacity had at this time been achieved by our ancestors.

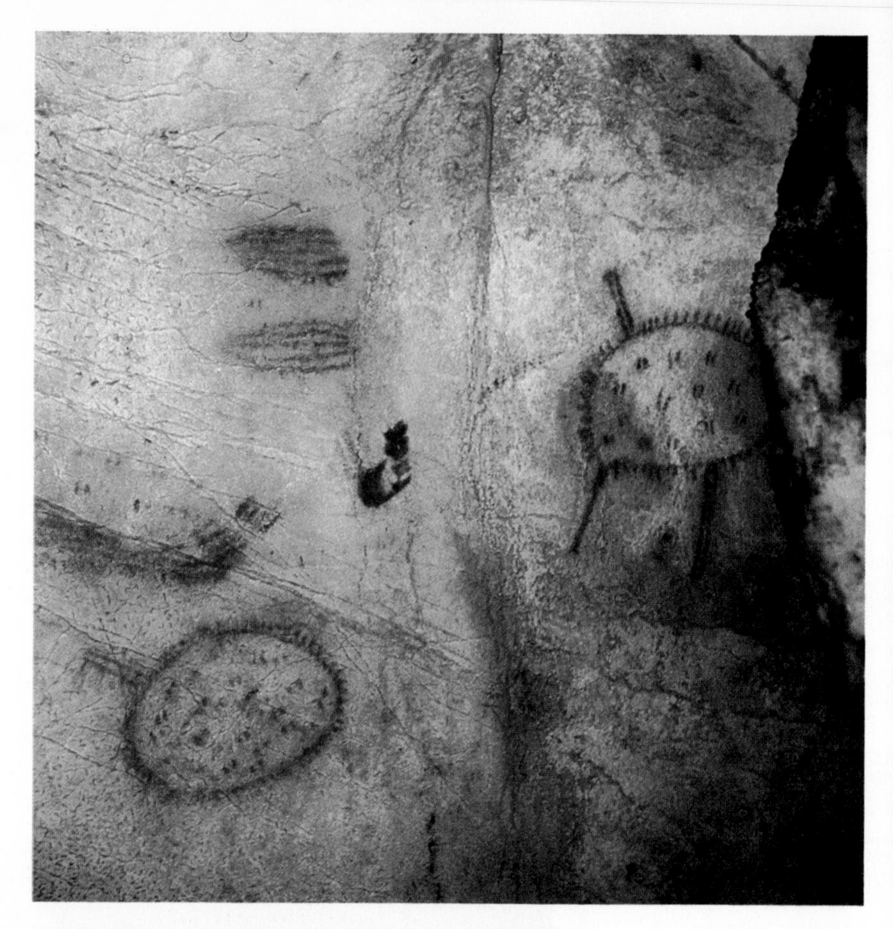

colourful and complex symbolism using wood, feather, ochre, blood, body-incisions, sand-drawings, songs, dances, myths and so on, none of which are readily preserved and so have little chance of entering into the fossil record. What one sees in the record must therefore be an impoverished representation of what actually occurred in the past. The samples of ochre that turn up in a number of sites throughout Europe, which are 200,000 years old or more, certainly suggest ritual adornment of people and their artifacts.

Ritual and symbolism hint strongly at linguistic competence. In considering the question of language among our forebears, Alexander Marshack suggests that although tool-making and a complex economy make enormous intellectual demands on people, they do not necessarily require the elaboration of a spoken language. 'Symbolic artifacts,' he argues by contrast, 'are not validated or explained by their utility in the natural or phenomenological world, even when they are derived from the real world or are directed at it. They are viable only within "artificial" contexts, and are possible only if they are named and if their use is defined. Language is therefore clearly necessary to such symbolic activities.'

A summary of the evidence

Given all these different strands of evidence what conclusions can we draw about when language emerged? In general, it would seem that the acquisition of spoken language during human evolution should be viewed as a stepwise process. In all likelihood, a rudimentary form of verbal communication arose as long as two million years ago, at the time of *Homo habilis*, and there may even have been language of some sort among the australopithecines. The emergence of *Homo erectus* was probably marked by a further development of this ability, with, perhaps, a greater vocabulary and a capacity for basic sentence structure. The evidence suggesting ritual acts in later *Homo erectus* populations might well indicate that there had been a further refinement of language to convey more subtle concepts. Finally, the symbolism and imagery embodied in the art that flourished from 30,000 years ago onwards surely signals the origin of modern language capacity, including the ability to articulate complex abstract ideas.

The purpose of language

If the question of *when* language arose is difficult enough, the issues of *how* and *why* are even more puzzling. Taking the *how* question first, one can propose that, as apes appear to demonstrate language competence, the evolutionary advance to full language capability is not particularly daunting. In other words, the progression from an ape-like ancestor who could make a limited array of pants, hoots and other sounds, to a hominid with the facility to name objects and construct meaningful sentences is well within the bounds of biological possibilities.

But some scientists see the large neurological differences between apes and men as indicating that there cannot have been a direct transition. Gordon Hewes of the University of Colorado has proposed that a stage involving an elaborate repertoire of gestures was crucial to the eventual emergence of spoken language. He points out that 'the initial appearance and early development of a gesture language did not require new anatomical structures or behaviour patterns'. Does the universal habit of gesturing during speech, especially when the speaker is searching for a suitable word, represent an echo of some early stage in human language development? Some researchers certainly view this piece of behaviour as being highly suggestive.

The gesture theory of language origins links it with the increasing complexity of stone-tool technology: the greater manual control being developed for manufacture of artifacts runs parallel with the increased dexterity in performing precise gestures. It is, however, possible to link rising manual dexterity with the emergence of language, without supposing that there was ever a 'gesture stage'. For instance, Ralph Holloway believes that 'tool-making and language skills are similar, if not identical, cognitive skills'. In other words, both processes involve organized, stepwise actions, and the imposition of arbitrary form or an arbitrary label on part of the environment. Ralph Holloway also reminds us that the area of the brain that governs fine actions of the hands and the area which controls the muscular movement required in speech lie very close to each other, a proximity that might well reflect shared elements in their origin.

Roger Fouts points to the tongue as the key organ in the origin of the spoken word: 'In order to produce speech you have to be able to stop your tongue at fifty different positions in your mouth. The stop-stop-stop process is very rapid of course, and although it may sound a daunting task, it is far less demanding neurologically than, say, the manual skills required in playing the piano. I see a very close evolutionary relationship between advanced control of the two things. Think of what happens to your tongue when you are trying to do a tricky manipulative task, such as threading a needle: the tongue pops out and moves in sympathy. And young children almost always protrude their tongue when they're writing.'

The question of *why* language evolved was once easily answered. It was believed that the organization required of a band of people involved in co-operative hunting demanded an efficient mode of communication, that is, speech. But in fact, hunters rarely talk when they are in search of prey, and the hunting dogs of Africa conduct highly complex and co-ordinated hunts without the benefit of speech as we know it. This hypothesis for the origin of speech must therefore be seen as far too simplistic.

As outlined in an earlier chapter, our ancestors moved from being opportunistic omnivores to operating a food-sharing economy based on meat and plant foods. This eventually led to the establishment of a hunting-and-gathering economy. It is surely possible that the practice of hunting-and-gathering could have been operated within a social system akin to that of modern apes, particularly chimpanzees. But as Glynn Isaac puts it: 'It is clear that the adaptive value of food-gathering and division of labour would be greatly enhanced by improvements in communication; specifically, the passage of information other than that relating to the emotions, becomes

A !Kung foraging band at their evening camp. Information is shared as to the whereabouts of herds of animals, or bushes full of ripe berries. Food is shared and plans are made for the following day. Holding this small community together is a network of close social interactions that could not be maintained without language.

highly adaptive. This has proved to be the case in other zoological phyla that have made the acquisition of food a collective responsibility, as is shown, for instance, by the development of the so-called language of the bees and other social insects.'

The emergent hominid way of life involved co-operation in food-gathering, systematic and reliable sharing of food, social life focusing on a series of temporary home bases and, probably, a division of labour. Language within *this* context is clearly so much more beneficial than it would be for more mundane tasks, such as passing on instructions for making stone tools or for planning a hunting expedition. Certainly, language would facilitate these activities, but they do not demand the spoken word in the same way that a co-operative economy and a complex social life do. In a small hunter-gatherer community social rules, elaborated through language, produce a cohesion that would be impossible to achieve in any other way.

Perhaps the most pervasive element of language is that, through communicating with others, not just about practical affairs, but about feelings, desires and fears, a 'shared consciousness' is created. And the elaboration of a formal mythology produces a shared consciousness on the scale of the community. Language is without doubt an enormously powerful force holding together the intense social network that characterizes human existence.

9 Neandertal Man

One afternoon in late September, some twenty years ago, Christos Sarianniddis, or Uncle Philipos, as he is called, led four of his friends to a spot on the slopes of the Katsika Mountain just above the village of Petralona, which is about 50 kilometres (30 miles) southeast of Thessaloniki. Uncle Philipos was a shepherd and for years he had thought there must be an underground cavern at this spot because he had noticed that in summer, cool air issued from a crevice, while in winter the same crevice seemed to be releasing warmer air. Philipos wanted help to widen the hole and explore the cave, thinking that perhaps they might find water which would help the village in times of drought.

After clearing away some rocks, the five men squeezed their way down into the ground. They descended about 5 metres (17 feet) and found themselves in a long, narrow cavern. But there was no underground river as they had hoped, and it was about a year before they returned. When they did visit the cavern again it was to look at a 'bear skull' that one of the men had seen while making the first exploration.

Quite unexpectedly they came across not a bear's skull but something that looked rather like the skull of a large monkey. It was sticking to the wall of the cave and it was partially covered by a thin coating of calcite crystals, a characteristic of these stalactite-filled limestone caves. The men were startled and decided to report their find to the scientists at the nearby Thessaloniki University. The 'monkey skull' turned out to be a fossilized cranium of an early human ancestor, and it is probably the best specimen of the people who inhabited Europe some 400,000 to 200,000 years ago.

Since this discovery, a great deal of work has been undertaken at the Petralona cave and a large number of stone artifacts and animal remains have been recovered. The cave itself has been beautifully lit and opened to visitors, and there is a fine local museum under construction.

The actual age of the Petralona skull is a matter for debate and Aris Poulianos, head of the Greek Anthropological Association, favours a much earlier age than the majority of scientists are prepared to accept. What is perhaps more important is the fact that the skull is remarkably complete and provides crucial information on what is probably the transition between the later types of *Homo erectus* and their successors in Europe, the Neandertals *Homo sapiens neanderthalensis*. The skull shows a mixture of features, a mosaic of some *Homo erectus* characteristics and some Neandertal characteristics. Other such skulls, showing a similar mosaic of features, have been found at various sites in Europe. These finds are important pieces of evidence in our attempt to understand the evolutionary relationships between fully modern man *Homo sapiens sapiens*, *Homo sapiens neanderthalensis* and *Homo erectus* in Europe and the Near East.

A group of men from the village of Petralona, above the cave where the 'monkey skull' was found twenty years ago. Uncle Philipos, the shepherd who first explored the cave, is standing second from the right.

More mosaic skulls from Europe

A series of lucky finds at a gravel pit in the village of Swanscombe, near London, produced part of a cranium which, like the Petralona individual, seems also to be at an evolutionary halfway house. The Swanscombe site had been known to be a rich source of prehistoric stone tools for a very long time, but it was not until 1935 that any human remains were found. Then, one Saturday in June, a workman finishing for the day noticed a piece of bone protruding from the gravel bank. His keen eye had spotted what turned out to be part of the back section of an ancient human skull.

The following March a second fragment was found, and this joined neatly with the first. Then, following a twenty-year break, during which time the area was extensively exploited for building material, a third piece was discovered. Unbelievably, it slotted together with the other two to form the entire back section of the cranium. This individual had lived at Swanscombe some 250,000 years ago, in a warm period between two Ice Ages. These periods are known as interglacials, and the climate then was substantially warmer than it is now. He would have seen elephants and rhinos in the Thames Valley forests, as well as wild boar and deer. As far as is possible to tell, the Swanscombe cranium once held a brain of about 1,300 cubic centimetres (78 cubic inches) capacity and the general shape was more rounded than the sharp, angular head typical of *Homo erectus*.

An interesting partner to the Swanscombe cranium comes from Steinheim in West Germany. Discovered just two years before the Swanscombe fossil, the Steinheim skull also came from gravel deposits and it too

The Petralona skull shows a curious mosaic of features, some typical of *Homo erectus*, others belonging to their European successors, the Neandertals. Other skulls, showing a similar mosaic of features, have been found in England, France and Germany.

is close to a quarter-of-a-million years old. Again, the brain is relatively large, and the cranium is well rounded. The face, however, is a surprise: it is very similar to *Homo erectus* in that it has prominent brow-ridges and a gently sloping forehead. Like the Petralona skull, that of Steinheim shows a curious mosaic of *erectus* and *sapiens* features.

Travelling now to a cave near the foothills of the French Pyrenees, which overlooks the prosperous vineyards of the Rousillon Plain, one finds yet another such skull. The Arago cave, fringed with fragrant herbs and perched precariously above a rock-filled gorge, is packed with thousands of stones and animal bones of many types: rhinoceros, elephant, musk ox, cave bear, giant sheep, lion, panther, beaver, and countless rodents. Among the fossil debris of some 400,000 years ago was the face and part of the skull of a young individual. Henri de Lumley, who with his wife Marie-Antoinette, supervises the excavation, speculates that the face might have been sliced from the skull to be used as some kind of mask in an ancient ritual. Whether or not that was the case, the fact that pollen from wild grapes was discovered within the skull may indicate that the French taste for the fruit of the vine began a very long time ago!

Once again, the Arago specimen clearly reflects its *Homo erectus* heritage, but shows some other features which look more advanced. The fossil record is regrettably sparse over this period of 400,000 to 200,000 years ago, but there are a number of other specimens that, like those I have mentioned, show this mosaic of features. These observations might suggest that this was a time of evolutionary change among our ancestors. Precisely what was happening, however, is unclear, as we shall see.

The point in question is the origin of modern humans, *Homo sapiens sapiens*. Three things can be said with some confidence. Firstly, from somewhere in excess of one-and-a-half million years ago to around 300,000 years ago *Homo erectus* existed throughout the Old World as a relatively stable species, and, as far as is possible to guess, was the rootstock from which modern humans eventually evolved. Secondly, the first fossil remains that are unquestionably fully modern man, *Homo sapiens sapiens*, appear in the fossil record at about 40,000 years ago. Thirdly, before this time and back to just over 100,000 years ago, there are specimens that are certainly *Homo sapiens* but with some primitive features. In Europe and the Near East these fossils are Neandertal Man or *Homo sapiens neanderthalensis*, while in Southeast Asia they have been called *Homo sapiens soloensis*. In Africa, an increasing number of finds have been made that are relevant to this crucial phase, but no sub-species name is in common use. Good examples are the famous Rhodesian Man from Broken Hill in Zambia, several skulls from Kenya and Tanzania and finds from Ethiopia.

The important question is whether or not it is possible to link *Homo erectus* with fully modern man through these more primitive *Homo sapiens* examples. In the case of Europe, did the Neandertals evolve from *Homo erectus* and then give issue to *Homo sapiens sapiens*? Or did Neandertal Man arise from *Homo erectus* only to become an evolutionary dead end, while modern humans evolved directly from *Homo erectus* independently of the Neandertals?

The discovery of Neandertal Man

Neandertal Man could not have been discovered at a less propitious moment than in 1856, when nineteenth-century society was violently opposed to any suggestion that humans were derived from the animal kingdom. His bones were unearthed from a small quarry in a cave near Düsseldorf. The cave was situated high up a small but steep-sided valley that carries the Neander River, a small tributary of the Rhine. Workmen were digging for

Richard Leakey beside the Neander River in northern Germany, with the skullcap and leg bone of the original Neandertal Man, discovered in 1856.

limestone in the cave floor when they came across some old bones. Bones, however, were of little importance to these particular quarrymen, and much of what was probably a complete skeleton was lost. Fortunately the skullcap, some ribs, part of a pelvis and some limb bones survived, and the quarry owner took them to a local scientist, Johann Carl Fuhlrott, to see what he could make of them.

Johann Fuhlrott recognized the bones as human, and reasoned that the thick low skull, the prominent brow-ridges and the stocky, curved limb bones indicated an individual of some antiquity. Seeking the advice of an anatomy expert, Hermann Schaaffhausen, Fuhlrott had his view confirmed that the Neandertal Man was of the 'most ancient races of man'. By 'ancient' the two men had in mind a matter of a few thousand years, somewhere in the first few pages of written history perhaps, not the tens of thousands of years we now know to be the truth.

This tentative statement about the finds must be seen as bold in view of the outraged reaction from the rest of the academic world. Various extraordinary explanations of the curious bones were put forward in an attempt to show that they did *not* belong to a normal member of a race ancestral to human beings. One German anatomist attributed the bowed legs to a life on horseback, and suggested that the man had been a Mongolian Cossack of the Russian cavalry which had chased Napoleon back across the Rhine in 1814. The Cossack, according to the inventive anatomist, had deserted his army and had crawled into the cave to die. Another anatomist saw indications of an 'Old Dutchman' in the low-vaulted skull. According to yet another scholar, the cave man had suffered from rickets which gave him bowed legs. The pain of the disease caused him habitually to furrow his brow, thus producing the enlarged brow-ridges.

On and on flowed the 'scientific' descriptions, with the Neandertal fossil generally considered as belonging to a savage foreign race or being the victim of some awful disease. The most destructive blow of all came from the renowned German anatomist-anthropologist, Rudolf Virchow, who dismissed any idea of antiquity and proclaimed that the skeleton displayed the unmistakable signs of childhood rickets followed by arthritis in old age. The argument was then over, for a while at least.

It is interesting to note that when, in 1868, the skulls and skeleton of Cro-Magnon Man were discovered in the Dordogne they were immediately accepted by the academic community. The evident signs of antiquity, in the form of primitive stone tools and associated bones of extinct animals, did not upset the scientists since the human fossils had perfectly respectable rounded skulls and no 'barbaric' facial features. Cro-Magnon Man could be accepted as a normal part of human history, whereas Neandertal Man could not because he did not live up to nineteenth-century standards of humanity. Cro-Magnon Man was in fact 35,000 years old, one of the oldest representatives of *Homo sapiens sapiens* in western Europe.

Decade by decade following the discovery of the controversial fossil bones from the Neander Valley, more and more similarly 'distorted' individuals kept turning up in various parts of Europe: in Germany, France, Belgium and Yugoslavia. Very soon the explanations that they were all simply diseased foreigners began to seem unlikely, and the Neandertal Men were accepted as being members of an ancient, barbarous race having little or nothing to do with the history of 'true' men. The prevailing prejudice was clearly reflected in the influential work of Marcellin Boule of the French National Museum of History.

It was to Marcellin Boule's anatomical expertise that the reconstruction of a virtually complete Neandertal skeleton was entrusted in 1908. The skeleton, that of an old man, came from a cave near the village of La Chapelle-aux-Saints in the Dordogne. The completeness of the fossil material offered an excellent opportunity for creating an accurate image of Neandertal Man.

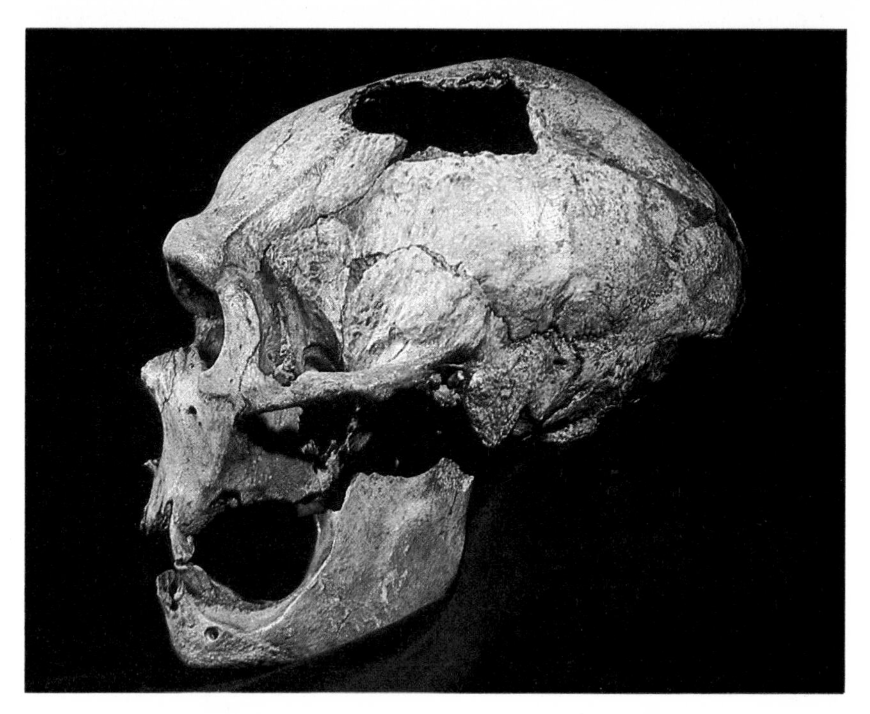

The skull of the old man from La Chapelle-aux-Saints, found in 1908. This skeleton was almost complete and Marcellin Boule used it to try to make an accurate reconstruction of Neandertal Man.

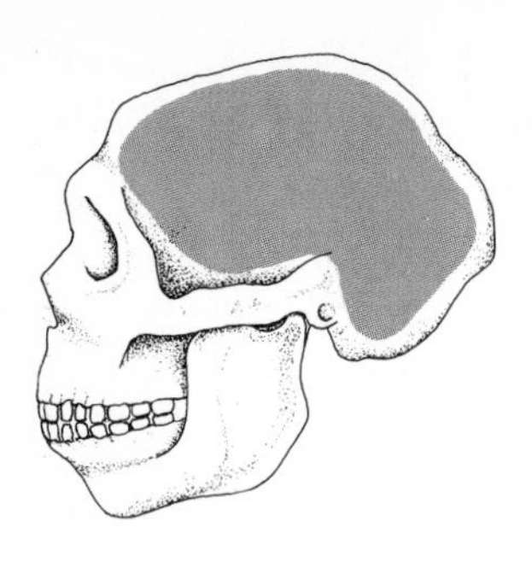

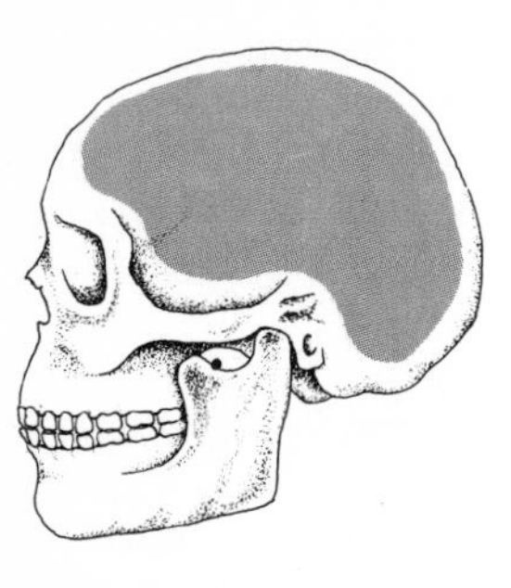

Above: Although the cranium of Neandertal Man was much lower vaulted than ours, it was also much longer, and the average size of the brain was, in fact, slightly greater than in modern humans.

Below: A reconstruction of Neandertal Man dating from 1929. The image of a brutish, dimwitted and round-shouldered creature persisted even as late as the 1950s.

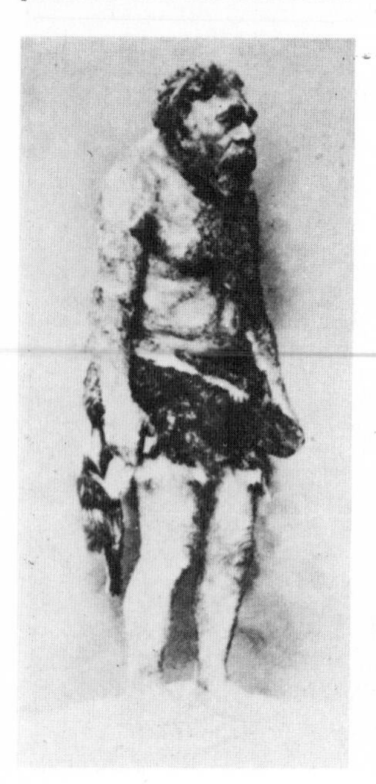

Yet, guided by his preconceptions, Marcellin Boule proceeded to emphasize everything that was primitive, brutish and ape-like about the skeleton. He even failed to take account of the fact that in this particular specimen the old man clearly *had* suffered from severe arthritis. Marcellin Boule's reconstruction stooped, with rounded shoulders and dangling arms. He walked on the outer part of his feet and with his knees bent. His big toes diverged from the rest of his toes, as in the apes, and his head was thrust forward in a cretinous and improbable manner. Despite the fact that the Chapelle skull had room for a brain bigger than a modern human's, Marcellin Boule deduced from the long, low shape of the skull that the old man had been dimwitted.

'The brutish appearance of this muscular and clumsy body, and of the heavy-jawed skull . . . declares the predominance of a purely vegetative or bestial kind over the functions of the mind . . . ,' wrote Marcellin Boule. 'What a contrast with the men of the next period, the men of the Cro-Magnon type, who had a more elegant body, a finer head, an upright and spacious brow, and who left behind so much evidence of their material skill, their artistic talent and religious preoccupations and their abstract faculties – and who were the first to merit the glorious title of *Homo sapiens*!'

Marcellin Boule published his conclusions in three large volumes between 1911 and 1913, and such was their academic weight that they determined the image of Neandertal Man for many decades to come. Admittedly, not all scholars took the one view, but when British prehistorian Grafton Elliot Smith wrote the following words in 1930, he was speaking for the majority: 'Neandertal Man is now revealed as an uncouth creature. . . . The hands are large and coarse, and lack the delicate play between thumb and fingers which is found in *Homo sapiens*. The large brain is singularly defective in the frontal region. It is clear that Neandertal limbs and brains were incapable of performing those delicately skilled movements that are the distinct prerogative of *Homo sapiens*, and one of the means by which the latter has learned by experiment to understand the world around him, and to acquire the high powers of discrimination that enabled him to compete successfully with the brutal strength of the Neandertal species.'

This view of Neandertal Man permeated textbooks and popular writing alike. The idea of a slouching, shuffling, inarticulate cretin was a powerful and persistent image; to refer to someone as 'Neandertal' was, and still is to some extent, an intended insult. Rehabilitation, however, began in the mid-1950s when two anatomists, William Straus and A J E Cave, undertook a second reconstruction of the old man of La Chapelle-aux-Saints. They recognized, as Marcellin Boule should have done, the distortion in the skeleton due to arthritis. Taking this into account, they were able to build a body which, though somewhat stocky, was essentially like modern man's. They wrote this of Neandertal Man: 'If he could be reincarnated and placed in a New York subway – provided he were bathed, shaved and dressed in modern clothing – it is doubtful whether he would attract any more attention than some of its other denizens.'

What, then, is meant, anatomically, by the term Neandertal? One is referring to an overall anatomical pattern, a collection of subtly different physical characteristics, rather than one particular feature. The cranium is relatively low, but not exceptionally so, and the brow-ridges are prominent. Both of these features are reminiscent of *Homo erectus* from whom the Neandertals were almost certainly derived. The brain is slightly larger than that of modern humans – 1,400 cubic centimetres (85 cubic inches), on average, as against 1,360 cubic centimetres (83 cubic inches) for people today. The limb bones are somewhat bowed, and the points of muscle attachment are strongly marked, both of which imply a very muscular build. Neandertals, both male and female, were probably much stronger than most modern people. The face is characteristic, with the nose and jaws thrust substantially forwards. The cheek arches slope backwards, and there is little forehead. The Neandertals' average height was around 1.67 metres (5 feet 8 inches).

The general impression, then, is of rather bulky, well built individuals. But whenever I pick up a Neandertal skull I am conscious of holding something that is unquestionably *Homo sapiens*. It is, however, a more primitive form of our species as is recognized in the scientific name *Homo sapiens neanderthalensis*, which distinguishes it from humans living today, all of whom are *Homo sapiens sapiens*.

The life of the Neandertals

Neandertal features first appear in the fossil record some 100,000 years ago, a time when the world was enjoying a warm interglacial phase that lasted from 130,000 years ago until around 70,000 years ago. The collection of characters seen in Neandertals became fully developed, however, during the last Ice Age, which began about 70,000 years ago, and the Neandertals are often thought of as the Ice Age people of Europe. This is clearly a simplification, firstly because they have their evolutionary roots in a warm era, and secondly because many Neandertals lived in areas of the world that were not

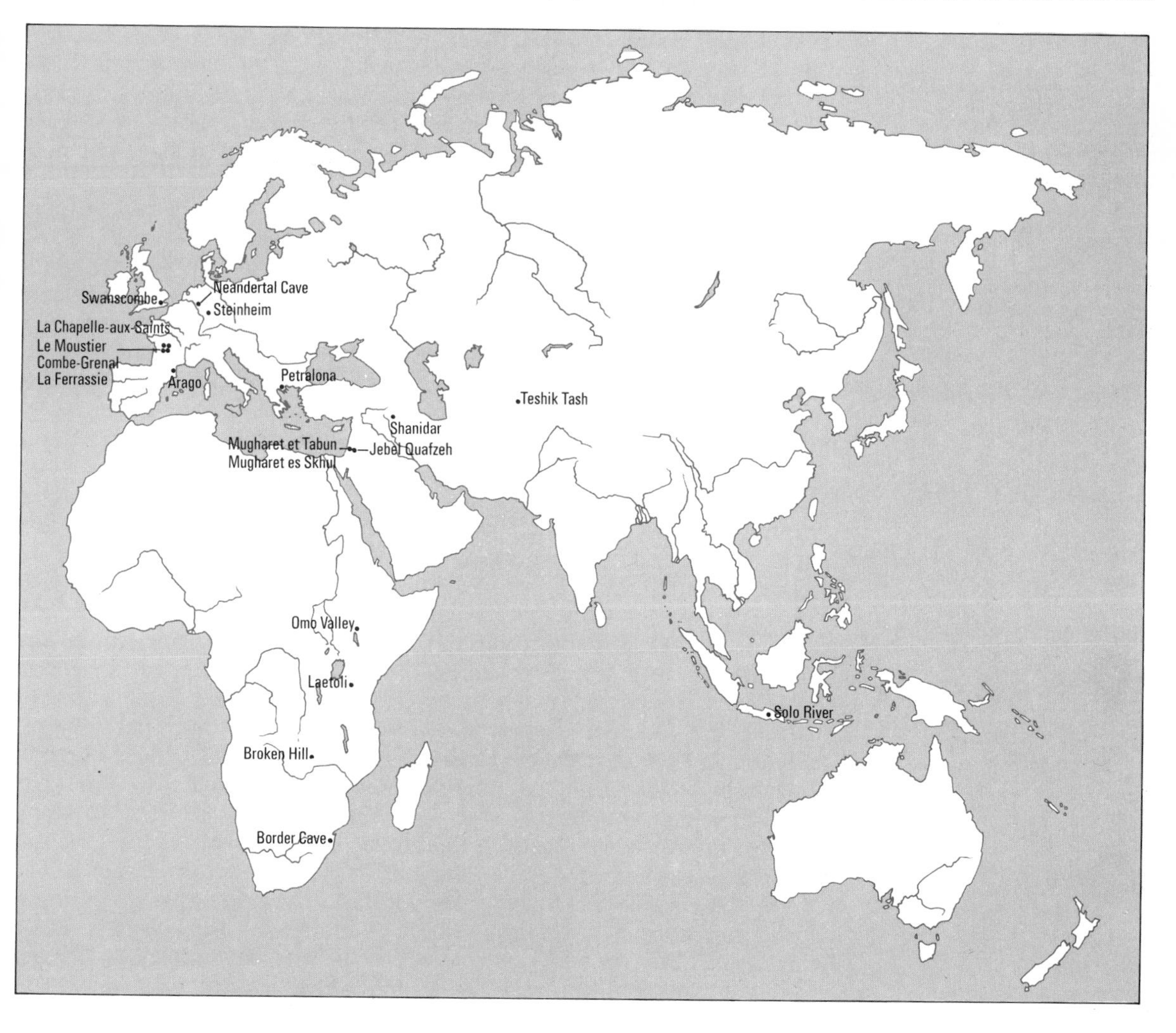

Some of the sites at which specimens of Neandertal Man and other early forms of *Homo sapiens* have been found.

in the grip of the Ice Age. Neandertal remains spread from western Europe, through part of the Near East and into western Asia. As far as one can infer from the fossil record so far, the Neandertals were the only types of humans to occupy this part of the world between 100,000 and 40,000 years ago.

How is one to explain the particular appearance of the Neandertal people? The stockiness of the body could be seen as a simple inheritance from *Homo erectus*, or alternatively as a specific adaptation to the environment. Bulky bodies and relatively short limbs are well suited to cold climates, as there is less surface area per unit volume from which to lose heat. In this sense, those Neandertals in the colder parts of their range were well adapted to coping with the low temperatures. Indeed, there is some evidence to suggest that Neandertals in western Europe had shorter limbs than those in the Near East where the effects of the Ice Age were less severe. It is of course possible to speculate that the ancestral *Homo erectus* population had itself undergone an anatomical adaptation to the cooler climes of northern Europe. In other words, European *Homo erectus* may have already been more stocky than populations in other parts of the world.

Protection against the cold has also been said to explain the Neandertals' projecting face: the greater separation between the nostril openings and the olfactory organs, which are located behind the bridge of the nose, it has been argued, would allow more time for the freezing air to warm, before striking the temperature-sensitive olfactory regions. However, Erik Trinkhaus of Harvard University suggests that much of the Neandertals' facial anatomy may be explained in terms of absorbing the enormous chewing stress imposed by the powerful jaws. 'But,' he cautions, 'there are almost certainly other factors involved too: cold climate may play some part, but we have to look for things that simply haven't been thought of yet.'

The evolution of *Homo sapiens* evidently involved a considerable expansion of the brain, and this is clearly manifested in their way of life. For a start, some of the Neandertal populations moved into parts of the globe where no hominid had ventured previously, and at a time when the intense cold of the Ice Age made life in those regions a considerable challenge. They were skilled hunters and were inventive in adversity. For instance, when there was little or no wood available, as was often the case on the treeless tundra, they used animal bones for building their shelters. They also burned bones as fuel in their hearths, which were constructed so that air would be drawn efficiently through them, thus promoting better combustion. Their camps were often quite large, and were sometimes occupied for many months. Clothing, of course, was essential to combat the freezing temperatures, and we may guess that their skill in making clothes increased considerably.

A new form of stone tool technology is closely identified with the Neandertals, called the Mousterian after Le Moustier, a cave in the Dordogne where it was first identified. The Mousterian technique was a considerable design improvement on its predecessor, the Acheulean. For instance, the Neandertals produced upwards of sixty identifiable items, including knives, scrapers and projectile points, each of which was prepared from a flake that was then trimmed for its specific purpose. The Mousterian tools were consistently finer and more precise than anything produced in previous cultures. A flaking technique known as the Levallois had been in existence for some time, but in the Neandertals' hands it was refined and developed to a high degree. Basically, it involved preparing a core so that many flakes of desired sizes could then be struck from it. These flakes formed the raw material from which the tool was shaped, by delicate and skilful trimming, a process that demanded fine control of the hands and

The Levallois technique of tool-making involved preparing a core from which a number of flakes of the right size could then be struck. A core is shown on the left, and a flake taken from it is on the right. This technique was far more economical in its use of stone than earlier methods.

a clear concept of the desired implement. This contrasts with the Acheulean industry in which each tool, in general, was manufactured from a single stone which was worked until it reached the desired shape. The Neandertals also made tools from bone to a much greater degree than their antecedents. Bone demands careful handling if useful tools are to emerge, but in skilful hands it is an excellent material from which to make fine and delicate implements.

By adopting the Levallois technique the Neandertals could generate as much as 2.2 metres (7 feet) of cutting edge from 1 kilogram (2.2 pounds) of flint: this is five times greater than that achieved by the basic Acheulean technique. Incidentally, successors of the Neandertal people invented a way of 'punching' fine blades from prepared flint blocks, giving them an incredible 26 metres (85 feet) of cutting edge from a single kilogram of starting material.

Although there is still some discussion about the matter, there appears to have been a number of distinct styles among Mousterian cultures, distinguished by the composition of the toolkit and by the form of the individual implements. François Bordes believes he has identified four such cultures in Europe, one of which, for instance, was characterized by tools trimmed so as to have distinct notches: this was known as the 'Denticulate Mousterian'. Others were termed 'Typical Mousterian', 'Acheulean type', and 'Charentian'.

François Bordes, who is a stone-knapper of world renown, envisages the coexistence of separate tribes of Neandertals, often living in close proximity but maintaining, over many thousands of years, their specific cultural identity. For example, people of the 'Typical Mousterian' culture lived at Combe-Grenal in the Dordogne, where, François Bordes believes, they concentrated on preying upon reindeer. Just 10 kilometres (6 miles) away lived another group of Neandertals, and these people were of the 'Denticulate' culture, for whom horses were apparently the major item of prey.

Many prehistorians have dismissed the notion of such strict cultural separation between people living close together. There are, however, parallels among contemporary people. British anthropologist Ian Hodder has been studying the lives of two groups of people, the Njemps and the Tugen, who live in the low rolling hills near Lake Baringo in northern Kenya. The two tribes have very similar economies, based principally on cattle, sheep and goats, they frequently trade with each other and even inter-marry. In spite of this they maintain a distinct identity in their style of dress, personal decoration, domestic architecture and technology. They also speak different, though related languages. When, say, a Tugen woman marries a Njemps man she moves to the neighbouring tribe and immediately adopts the local style of dress, decoration and domestic technology. Such strongly impressed tribal styles are, suggests Ian Hodder, an essential element in maintaining social and economic cohesion in complex communities. A particular way of doing things is a statement of belonging to a particular group. Did the 'Denticulate' and 'Typical' Mousterians of the Dordogne have such a relationship? It is certainly possible, because, from all that is known of the Neandertals, it is clear that they were very sensitive and socially developed people.

However, American anthropologists Sally and Lewis Binford propose a different explanation for the separate Mousterian industries. They suggest, very simply, that the different toolkits were designed for different tasks. One set could be specifically suited to woodworking, another to the preparation of hide, another to butchery, and so on. In many ways the simplicity of this functional explanation is attractive, but so far analysis of the association of tool industries with different types of sites has failed to produce any evidence in favour of the idea. The theory could, of course, be tested by using Larry Keeley's technique of microscopic examination of the tools, but such work has not yet been done. François Bordes's tribal explanation has its problems too, not least of which is that in a number of instances several cultures appear to have occupied the same locality at approximately the same time. To counter this it could be argued that, although the archaeological distance between two sets of industries at one site is small, the actual time separating the different occupations might have been many years. There might well have been some geographical drift of tribes over several generations, with one tribe coming to occupy sites previously belonging to another tribe.

This debate remains unresolved, but whatever the outcome it is evident that the Neandertal people possessed a considerable degree of technological skill and probably understood the world around them very well indeed. Without a keen sensitivity to the possible resources in the environment, combined with a consummate ability to exploit them, it would surely have been impossible for the Neandertals to have conquered such a diversity of environments. Anatomically, Neandertal Man has close affinities with modern humans. His technology is well developed and elaborate. There are also suggestions, in the different styles of the Mousterian culture, that there may have been separate tribes each with a well developed cultural identity. What, if anything, can one say about the spiritual aspects of Neandertal life?

Ritual burials

With the arrival of the Neandertals we find the first archaeological indications of ritual burial. Throughout the lands of the Neandertals there are the remains of individuals, young and old, who have been laid to rest in a deliberate and perhaps reverential manner. At Le Moustier a young teenage boy was lowered into a pit, to lie on his right side, his head resting on his forearm as if asleep. A pile of flints served as a pillow, and an exquisitely worked stone axe lay near his hand. The bones of wild cattle lay scattered around him, suggesting perhaps that meat was buried with him to provide sustenance for the journey to a new life. At Teshik Tash, in Uzbekistan, Central Asia, the skeleton of a young child lies among ibex bones, with six pairs of ibex horns forming a ring around his head. The child's skeleton also bears the marks of stone tools, suggesting that these were used to cut away his flesh, presumably as part of a ritual. And at La Ferrassie rock shelter, near the town of Le Bugue in the Dordogne, the headless body of a young child was found lying in a flexed position at the bottom of a shallow pit. Slightly higher up the pit is the jawless skull of a child lying below a limestone slab. The undersurface of the slab had been coated with red ochre, and the top is marked with eighteen small pits. These strange burials were undoubtedly of great significance to the Neandertals, although it is unlikely that we will ever understand their meaning.

One discovery that does strike a chord with modern man comes from the Shanidar cave in the Zagros Mountains of Iraq. The cave, which Ralph Solecki of Columbia University, New York, has been excavating for more than twenty years, has yielded a rich selection of Neandertal remains. One of these, Shanidar IV, was the subject of an unusual burial, one early June day, some 60,000 years ago.

Oak, pine, juniper and ash groves nestled in the numerous valleys around the Shanidar cave, and many wild flowers grew among the grasses. This is known from the work of Arlette Leroi-Gourhan, of the Musée de l'Homme in Paris, who has detected the pollen grains of these trees and plants scattered through the soil deposits of the cave. Their pollen would have drifted into the cave on the breeze, or may have been carried in by the inhabitants' clothes and bodies. Arlette Leroi-Gourhan was surprised, however, when she discovered dense clusters of pollen that could only have come from whole flowers. The arrangement of the flowers was not random: they were carefully placed around the body of the Shanidar IV man.

The pollen analysis revealed the presence of yarrow, cornflower, St Barnaby's thistle, ragwort, grape hyacinth, hollyhock and woody horsetail. The effect would have been a delicate mixture of white, yellow and blue flowers with the green branches of the woody horsetail which, according to Arlette Leroi-Gourhan, would have made 'a sort of bedding on which the dead could have been laid'. The scene is meaningful enough, but Ralph Solecki suggests that there might be more significance to the flowers. 'Most of them,' he points out, 'are known to have herbal properties and are used by the people of the region today. . . . One may speculate that Shanidar IV was not only a very important man, a leader, but also may have been a kind of medicine man or shaman in his group. . . .'

The import of the Shanidar events cannot be doubted, and, together with the many other examples of ritual burial, they speak clearly of a deep feeling for the spiritual quality of life. A concern for the fate of the human soul is universal in human societies today, and it was evidently a theme of Neandertal society too. There is also reason to believe that the Neandertals cared for the old and the sick of their group. A number of individuals buried at the Shanidar cave, for instance, showed signs of injury during life, and in one case a man was clearly severely crippled, probably being almost completely paralysed down his right side. These people lived for a long time although they must have needed constant support and care to do so. It is impossible to imagine a society as complex as the Neandertals' obviously was, and as imbued with tender feelings and manifestations of ritual, without a sophisticated spoken language.

Overleaf: An artist's impression of the burial at Shanidar. The ceremony that took place here strikes a chord with our own customs and beliefs, but the ritual burials of the Neandertals took many different, and often bizarre, forms.

The disappearance of the Neandertals

The Neandertal people no longer exist. What became of them? Although their entrance onto the stage of human evolution was relatively gradual, their exit appears to have been swift. In the east they vanished around 40,000 years ago, whereas in western Europe their disappearance came some five thousand years later. Did the Neandertals evolve directly into *Homo sapiens sapiens*, or were they replaced by modern humans who had evolved elsewhere on the globe? The first of these possibilities is suggested by what is known as the 'Neandertal phase' theory while the second is represented by the 'Garden of Eden' theory.

If one views the evolution of the *Homo* line as having more to do with the progress of cultural capabilities than with environmental conditions, then it is possible to imagine that *Homo erectus* populations throughout the world became more and more dependent on the development and exploitation of technology, and that this created its own selection pressure that propelled the species towards *Homo sapiens*. In each part of the world where there had been *Homo erectus*, there would eventually arise an early grade of *Homo sapiens*. As selection pressure continued through the demands of culture, each population of early *Homo sapiens* ultimately emerged as *Homo sapiens sapiens*, modern man.

In this model Neandertal Man is seen as the European and Near Eastern version of early *Homo sapiens*, hence the term 'Neandertal phase' theory. It is envisaged that there were other populations of early *Homo sapiens* in Africa and Asia which gave rise to modern man in those regions, but that these populations are not as apparent in the fossil record as the Neandertals, largely because very little detailed work has been done on this period in Africa or in Asia. The diversity of modern people is therefore explained to some extent in terms of their separate geographical origins, even though all humans belong to *Homo sapiens sapiens*. This model therefore presents the process of human evolution as a relatively smooth progression from one stage to the next that took place on a global front.

The 'Garden of Eden' theory, by contrast, proposes that the important steps in human evolution took place in restricted locations rather than on a worldwide scale. Imagine a small, isolated population of *Homo habilis* somewhere in Africa. Through genetic change it gives rise to *Homo erectus*. As *Homo erectus* is better fitted to the environment than is *Homo habilis*, its population rapidly expands, soon to cover the African continent and eventually to move into the rest of the Old World too. With such a wide geographical distribution, a certain degree of localized adaptation, such as slender tropical forms and more stocky temperate-region forms, inevitably develops: the fossil record does display a great deal of worldwide variation in *Homo erectus*.

More evolutionary advances occur among certain populations of the world's dispersed *Homo erectus* stock giving rise in some parts of Europe to the Neandertals, and in Africa to a different version of early *Homo sapiens*. The initially small population of Neandertals would again expand to take over territories previously occupied by *Homo erectus*, and the same would happen with the early *Homo sapiens* population in Africa. One *Homo sapiens* population (possibly the African one) then gives rise to *Homo sapiens sapiens*, and this new, highly successful species spreads throughout the Old World replacing all other *Homo* species including Neandertal Man. The diversity between today's populations is then explained simply in terms of 'skin deep' adaptations to local geographical conditions.

A theoretical objection to the 'Neandertal phase' theory is that, among other animals, it is known that new species *do* generally arise in small isolated populations. To propose that it is otherwise for humans is, therefore, special pleading. Countering this, one can argue that although the early human species were widespread they were very mobile and freely interbreeding populations. Under these circumstances, speciation may happen rather differently.

These, then, are the theoretical considerations over which biologists are currently arguing. What do the fossils tell us? As I indicated at the beginning of this chapter, certain European fossils, from Petralona, Arago, Swanscombe and Steinheim, give an impression that they are on the way from one stage of the human line to another, that they are 'in transition' to the Neandertal stock. If indeed these individuals did display a mosaic of characters ancient and modern, then this could be taken as support for the 'Neandertal phase' model, for it shows that there was gradual change over a broad geographical region. Is there similar evidence for a transition from *Homo sapiens neanderthalensis* to *Homo sapiens sapiens*?

To answer this question we go to the slopes of Mount Carmel overlooking the Mediterranean, near Haifa. There, in the late 1920s and early 1930s a joint Anglo–American team explored many caves and came up with some intriguing finds. At one, Mugharet et Tabun (Cave of the Oven), they discovered two skeletons of the classic Neandertal form, dating from at least 45,000 years ago. A few hundred metres away, at Mugharet es Skhul (Cave of the Goat Kids), they unearthed the remains of ten individuals that were much more difficult to classify. They were certainly not modern, in the full *Homo sapiens sapiens* sense, but neither were they archaic: they were

Looking down into the Mount Carmel cave known as Mugharet et Tabun: the step-like platforms are the legacy of archaeologists who first worked here during the late 1920s and early 1930s. They discovered two skeletons showing typical Neandertal features which were 45,000 years old.

something of a mixture, neither Neandertal nor fully modern man. These individuals, as far as can be ascertained at the moment, are some 40,000 years old, possibly just 5,000 years younger than their Neandertal neighbours at Tabun.

The Skhul individuals were not unique it seems. Farther inland from the Carmel range, and located close to Nazareth at the narrowest section of a pass through the mountains of Lebanon, is the large cave, Jebel Qafzeh. During a forty-year period of excavation which began in 1933, the remains of eleven modern-looking, though rather robust, individuals were found. These too are up to 40,000 years old. Despite their modern appearance, the Qafzeh people apparently made and used typically Mousterian tools.

It is tempting to see the Skhul and Qafzeh people as the type of intermediate one would expect if the 'Neandertal phase' theory were correct, and indeed many scientists interpret the evidence in exactly this way. William Howells and Erik Trinkhaus, however, do not agree. They argue that, in spite of the robust features of the fossils, statistical analyses of the overall anatomical patterns 'indicate that the specimens lack any particularly close Neandertal affinity . . . [and they] show no convincing sign of a total morphology that is transitional between Neandertals and modern men'. So, here once again we have a situation in which the same set of fossils are interpreted as implying exactly opposite conclusions by different groups of scientists.

What, then, of the alternative: that modern man evolved elsewhere in the world, possibly in Africa, and then spread into Asia and Europe? Sadly, there is no good fossil evidence that swings the debate firmly in this direction either. There is a very modern-looking cranial fragment and lower

The entrance to Mugharet es Skhul, only a few hundred metres from Mugharet et Tabun. Here the finds represented ten individuals which could not be classified as true Neandertals nor as fully modern man. These skeletons, with their strange mixture of Neandertal and modern features, are only 5,000 years younger than the people of the Tabun cave.

jaw from deposits in Border Cave, South Africa, which may turn out to be as old as 90,000 years, but which could be considerably younger. If the older date were to be correct then the 'Garden of Eden' theory would certainly gain weight from it. And a 120,000-year-old skull from Laetoli has a number of tantalizingly modern features, as do two skulls that I collected in the Omo Valley. These again turn the focus to Africa. If the 'Garden of Eden' theory is correct, then an African origin of modern man would fit certain of the facts as they appear at the moment, but some would say that the Middle East has an equally strong claim. It simply is not possible to resolve the question on the information currently available.

Regarding the debate between the 'Neandertal phase' theory and the 'Garden of Eden' theory I believe that the answer may eventually be shown to lie somewhere between the two extreme theoretical positions. There may well have been an evolutionary spurt in some localized population, giving rise to *Homo sapiens sapiens* who then spread out into other parts of the world. But instead of totally replacing native populations of other *Homo* species, it seems likely that the genetic difference between the new and the old would be small enough for a good deal of interbreeding to be possible. None of the Neandertal sites known reveal any signs of a bloody takeover by invading hordes. Indeed, what relevant evidence there is of this period of change indicates a smooth transition. So, rather than replacement of indigenous people, there was probably an assimilation through interbreeding. The numbers of people moving into Europe, the Near East and western Asia may well have been so large by comparison with the resident Neandertal population as to dilute the effect of the Neandertal genes quite considerably. This would explain the rapid disappearance of the Neandertals' characteristic features from the fossil record. Modern people of this part of the world should therefore think it likely that they have inherited at least a few Neandertal genes.

Following the development of fully modern man, the tundras of northern Europe and Asia were penetrated, and from there the New World was entered at last. The ice sheets 'locked up' a great deal of water, so that the sea level fell, and the journey from Southeast Asia to Australia involved just a few island hops rather than the major sea venture it would involve today. The challenge for the first Americans was, by contrast, a long migration over frozen tundra and grassy plains across the temporarily dry Bering Straits, a journey that appears to have first been undertaken around 40,000 years ago. Very probably, these people were simply following herds of prey animals rather than embarking on a deliberate migration. Once in the New World, however, they began to colonize new territories, grasp new opportunities, and overcome new challenges.

But mankind was not only expanding in the geographical sense. It is clear that there was a distinct leap in manipulative and organizational abilities following the appearance of *Homo sapiens sapiens*. Accompanying this was a sudden rich efflorescence of art: cave painting, carving and engraving become an important and integral part of human life. The Age of Art had arrived.

10 Ice Age Art

One of the most exciting experiences of my life was to be confronted by the fantastic paintings on the walls of Lascaux cave in southwest France. The vivid forms of horses, stags and oxen seem to leap out from the glistening crystalline surface on which they were painted some 14,000 years ago. These are not careful portraits of quiescent animals. They are bold images filled with action, movement and life.

As if fleeing from a strange, horned, sorcerer-like figure poised close to the entrance, the cavalcade of prehistoric beasts stampedes towards the deeper recesses of the cave. Four gigantic white bulls in black outline dominate the long cavern where it widens to form a rotunda: this is called the Hall of Bulls. A host of smaller creatures jostle among the legs of the great beasts. Trotting horses, tense stags and frisky young ponies stand out from the walls and ceiling in black, red and yellow, sometimes bold, sometimes just tantalizing suggestions. Some images obscure others; some are huge; some are diminutive. A fine purple-red horse with a rich flowing black mane hangs near two great bulls facing each other in head-on challenge. Geometric signs and rows of black dots add to the mystery of the Hall of Bulls.

The cave narrows again beyond the rotunda, creating a gallery of leaping and falling figures. The gallery opens with the boldly sketched head of a magnificent stag. A black cow jumps across the ceiling from one side of the alley to the other. On the right wall, beneath the enormous head of a black bull, is a file of small, brown, long-haired ponies. A herd of thirteen horses stands close by. The opposite wall shows a black bull racing towards the end of the gallery, and before it flees a horse, its black mane flying in the wind. As the gallery narrows still further and twists sharply to the right into darkness, another horse plunges and tumbles through space.

From the Hall of Bulls a small exit to the left leads into a narrow corridor whose crumbling walls are etched with a tangle of tiny engravings, seen at their best by an angled light. Miniature horses and stags abound, sometimes the whole animal, sometimes just the head. A small swelling in the rock face seems to have been used to portray a rounded belly, and one tiny protuberance forms the eye of a horse. The corridor opens into the Nave which displays four groups of paintings, three to the left and one to the right. Eight wild goats cluster on the left. Four are red with black horns, while four are black with horns seen only in engraving, the colour having faded long ago. More geometric designs follow, together with two pregnant mares. A stallion and a bison are pierced by engraved arrows. Two more horses are depicted, one galloping, the other grazing. Most curious of all on this wall is a vast black cow painted over a series of much smaller horses. The animal's gigantic body is supported on slender legs and is topped by a tiny head: it is quite unlike anything else in Lascaux. Gliding

A magnificently painted bull from the cave of Lascaux in southwest France.

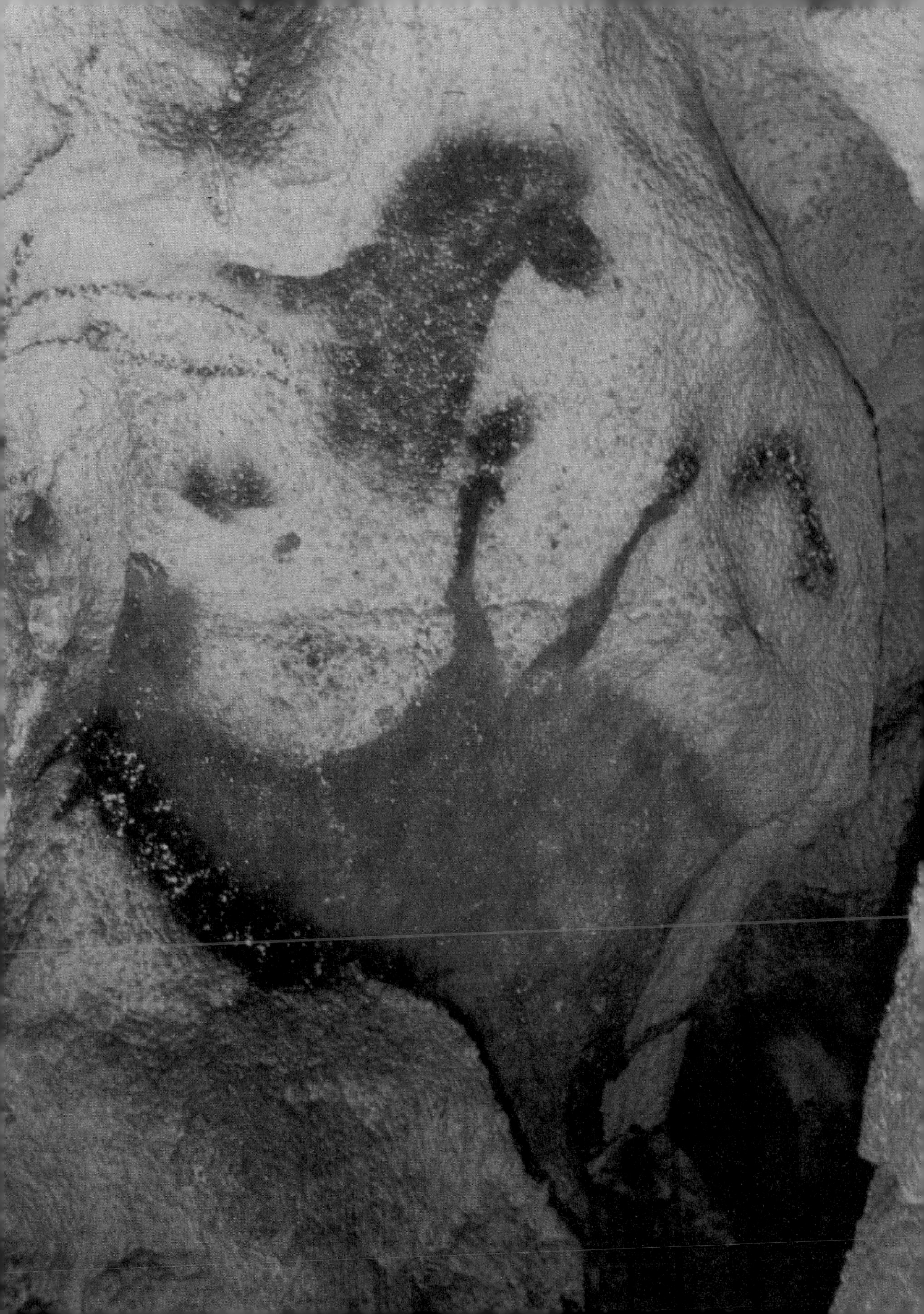

At the far end of the Hall of Bulls, a horse appears to tumble backwards into the darkness beyond.

silently along the opposite wall are five stags sketched in simple black strokes. Only their heads and necks are visible, as if they were swimming across a river, and the leading animal seems to be raising its snout as it nears the invisible bank.

The Nave narrows at its end, then widens to form two recesses, one of which has six lions engraved on the walls. One lion has been killed and twelve arrows protrude from its finely shaped body. Opening to the right at the junction between the engraved corridor and the Nave is the Apse, an area laced with engravings and many faded paintings. It is difficult to make out many clear figures here, but one large engraving of the head and antlers of a stag is incredible. It is surely one of the finest pieces of engraving in prehistory.

The mystery of Lascaux deepens with a curious scene depicted on the wall of a well in the Apse. A man lies dead between the figures of a wounded bison and a rhinoceros. Unlike the animals in the cave, the man is sketched in stick form. He has four fingers at the ends of his matchstick arms, and his face looks like a beak. Near him lies a pole with a bird perched at the top, though whether this is meant to be a real bird or a carving is impossible to tell. The bison is losing entrails from its wounds, its hair bristles and its tail is raised as it threatens the man with lowered horns. Three pairs of dots separate this scene from the rhinoceros which is facing the other way and apparently departing.

You emerge from Lascaux cave, dazzled by the brightness of the wooded glade that fringes the cave entrance and knowing that you have left behind an ancient and extraordinary world. An overwhelming sense of pulsating activity pervades the uncanny quiet and stillness of the cave. The contrast between the vital animals on the walls and the absolute tranquillity of the cool cave air is unnerving. One marvels at the artistic skill shown in those ancient paintings. But most of all, one's mind is sent whirling back through time, as one wonders about the people who made the paintings. What motivated them? Was the cave a sacred or a holy place? Were scenes of hunting magic played out there? Were acts of social or seasonal ritual enacted in front of the freshly created images? Or did these people simply glory in the sensual pleasure of their artistic creations?

Today, two sets of huge doors at the cave entrance separate the modern world from the now silent world of the prehistoric artists. These are a protection against vandals and against the entry of bacteria and fungi that could destroy the precious paintings. The age of prehistoric art flowered towards the end of the last Ice Age, beginning about 35,000 years ago, reaching its height at the time of Lascaux some 15,000 years ago, and fading forever as the ice sheets retreated 10,000 years ago. So far, over 200 decorated caves and shelters have been discovered, principally in south-west France and northern Spain. Lascaux and the Spanish site of Altamira, the first painted cave ever to come to light, are undoubtedly our most splendid legacies of prehistoric art in Europe.

There are also some spectacular cave paintings from African sites although these are less known and often disregarded. These sites fall into two main areas: the North African sites and those south of the present Sahara desert, in Tanzania and South Africa. The Tanzanian examples are the most interesting and my parents worked on these in the early 1950s, my mother producing a most detailed record by carefully tracing the images at some of the eighty-eight recorded sites.

The paintings in Tanzania are on the exposed walls of rock shelters and are therefore poorly preserved. However, excavations of several sites show that the paintings were almost certainly made from about 35,000 years ago onwards. The paintings include a large number of magnificent animals that were shown in superb detail as well as some wonderful depictions of humans. Often humans are shown in some activity – hunting, bathing and, in one case, arguing over an individual who is thought to be a young maiden.

Europe in the Ice Age

Ice Age Europe would have looked totally different to us, both in physical aspect and in the animals that lived there. Glaciers covered Scandinavia and most of Britain, and to the south of these treeless tundra and steppe stretched endlessly, populated by huge herds of reindeer, mammoths and horses. In southern France and Spain the climate, though cold, was less gruelling: summer temperatures averaged 15°C (59°F), with frosty evenings. Cliffs in the limestone hills offered shelter from the worst rigours of winter. Mammoth and reindeer lived in the south during the colder spells, and there were herds of bison, deer and wild oxen.

The combination of hills, river valleys and plateaus in France and Spain allowed a great variety of animals to survive there, each adapted to particular environmental conditions. The woolly rhinoceros and the musk ox, like the mammoth and reindeer, were best suited to colder regions. Ibex and chamois roamed the rocky slopes of the ice-capped hills. Wild boar, deer and Irish elk sought the shelter of the forests in the lower mountain slopes. The rivers were rich in salmon, pike, trout and eel. Lions, leopards, wolves and foxes were among the carnivores who preyed on the vast herds.

Ancient rock paintings from Africa, unlike those of the European caves, often depict human figures engaged in some activity. This painting from a rock shelter in Tanzania is believed to show two groups of people arguing over a young maiden.

Southern Europe, during the Ice Age, was immensely rich in wildlife, and yet the human population was sparse, numbering perhaps a few tens of thousands. François Bordes once referred to it as 'a human desert swarming with game'.

For the most part, the nomadic hunter-gatherers of the time lived out in the open, erecting temporary shelters in favoured locations. Sometimes they lived in caves as the litter of domestic refuse shows. But it is the paintings and engravings in the caves that interest us here, and generally it seems that the parts of caves that were decorated did not serve as living quarters. It is possible that paintings were also made on the walls of the cave mouths that served as shelters, but, because these shelters are relatively exposed to the elements, their paintings gradually disappeared. However, it is interesting that there are many examples of paintings, engravings and other artistic work located in places to which access was extremely difficult and which were probably visited only once or twice as shown by the presence of undisturbed footprints. The art of these caves would appear to have been something rather special. Moreover, very few of the more accessible decorated caves have a significant amount of food litter in them, which again indicates a special function for the sites.

The finding of the first cave

The old farm of Altamira, meaning 'high lookout', is situated on gently sloping but high meadowland some 4 kilometres (2·5 miles) from the coast of northern Spain. To the south, the Cordillera Cantabrica dominates the skyline, and the frequently snowclad Picos de Europa rise to peaks of almost 3,000 metres (10,000 feet) in the west. It is a dramatic setting. Far underground, beneath the farm, a series of caverns and narrow corridors snake through the limestone. That the area is riddled with caves is common knowledge, but until 1868 the now famous Altamira cave was unknown to the owner of the land, Don Marcellino de Sautuola. In that year a hunter came across the cave entrance as he attempted to rescue his dog who had fallen among some rocks while chasing a fox.

When Sautuola heard about the cave under his farm he was interested enough to explore it briefly. He was something of an amateur archaeologist, but apart from a few old bones he saw nothing unusual. While he was in Paris in 1878, Sautuola talked to the famous French prehistorian Edouard Piette about life in the Ice Ages, and Piette told Sautuola what he might look for in the cave. Inspired by his discussions, Sautuola reopened the Altamira cave and explored it more thoroughly. He found that the small entrance widened to a zig-zag chain of three galleries with several side extensions. Eventually the galleries tapered to a long, narrow, twisting corridor extending some 50 metres (165 feet), making the cavern system some 300 metres (1,000 feet) in total.

Working carefully on his hands and knees, Sautuola discovered a number of stone tools, but little else, and the cave's rich secrets might have remained hidden forever were it not for his young daughter, Maria. Accompanying her father one day in 1879, Maria wandered into a low chamber which Sautuola had explored previously. Whereas her father had been forced to crawl through the chamber Maria could stand upright in it, and, looking up, she caught a glimpse of coloured images painted on the low ceiling which her father had not noticed. Sautuola could hardly believe what he saw when Maria called him to the chamber. The red figures of almost two dozen bison hung there in the flickering lamp light. Other animals were depicted on the periphery of the group: two horses, a wolf, three boars and three female deer.

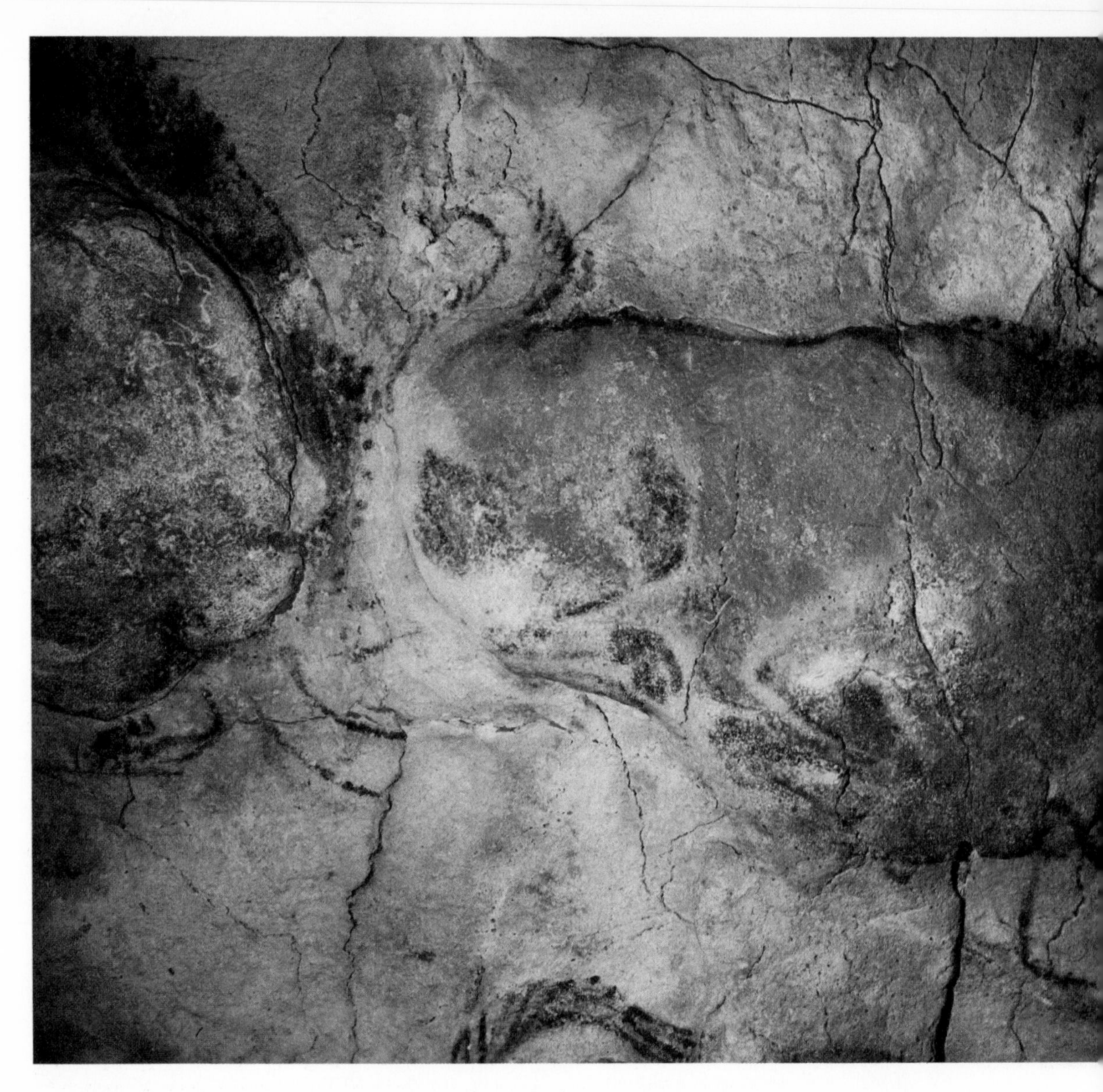

One of the bulls from the cave of Altamira; the hindquarters of another bull can be seen on the left. Natural bumps in the ceiling of the cave have been used to give the paintings a three-dimensional quality. The lowered heads and curved backs of the bison have been interpreted in various different ways: some experts believe the animals are charging, others think they are in the throes of death, or giving birth.

Coloured in red, yellow and black, the scene was as fresh as if it had just been completed. In ingenious ways, the Palaeolithic artists had exploited the humps and hollows of the undulating ceiling to give a remarkable three-dimensional quality to the pictures. This is a common feature of Ice Age art, but nowhere is it done so well as at Altamira.

When Sautuola had visited Paris in 1878, he had seen at the Grand International Exhibition a display of engraved stones collected from a number of French caves which were accepted by the academic community as being prehistoric. Sautuola saw in the Altamira paintings echoes of the images in those engravings. His excitement and delight can be imagined, and so too can his shock and disappointment when the scholars of Europe rejected the paintings as being nothing more than the work of a modern artist. One Spanish expert stated that the paintings had 'none of the character of either Stone Age, archaic, Assyrian or Phoenician art. They are simply an expression of a mediocre student of the modern school.'

And a French scholar even went so far as to point an accusing finger at an artist called Retier who had stayed with Sautuola for some time. Hurt and upset by this callous treatment from the scholars, Sautuola had the cave closed, and he died in 1888 with his discovery still unacknowledged. Ironically, Europe's scholars could not accept the authenticity of the Altamira paintings precisely because of the supreme skill they embodied.

Not all the scholars were set against Sautuola, however. Edouard Piette possessed sufficient vision to see Altamira as the product of the Ice Age. A year before Sautuola died Piette wrote to Emile Cartailhac, the leader of the opposition to Altamira's authenticity, encouraging him to reconsider his position. His plea failed and Altamira remained unimportant to the academic world for more than two decades following its discovery. But, as with Neandertal Man, more and more similar finds were made and this finally forced a reassessment. First, La Mouthe, a cave with painted and engraved bison and a fine example of a stone lamp, was discovered in the Dordogne in 1895 and the dating of this to the Ice Age was incontrovertible. Then more examples of decorated caves came to light in France: Font-de-Gaume and Les Combarelles in the Dordogne, for instance. Opinions began to sway, and the final turning point came in 1902 when Emile Cartailhac accepted his mistake and announced it to the world in an essay entitled *Mea Culpa d'un Sceptique*. Altamira was accepted as authentic.

Many of the 200 or so decorated caves and shelters now known were found by accident, with dogs, children and pot-holers often playing a part in their discovery. These evocative legacies from the Ice Age are concentrated in the limestone hills and valleys of the Périgord, the Pyrenees and Spanish Cantabria, with a few examples in Italy and a solitary site from Kapovaia in the Urals. There is a striking similarity among the wall paintings throughout these regions, not just in the subjects, but in elements of style too. Each cave and shelter is unquestionably unique in composition and may also display special technical innovations, but overall they give the impression of variations upon a theme.

During my visits to a number of the French sites I was struck by how well the artists knew their subjects and their fine sensitivity for detail. Nevertheless, the quality of the art does vary considerably, with Lascaux standing out as being far superior to most of the others. Much of the cave art projects a suggestion of chaos: large images are jumbled with small ones, and the painting of a second picture over an earlier one is frequent; there is no obvious boundary to the work, animals may be oriented in any direction, there is no context in the form of landscape or even a surface for the animals to stand on. With the exception of Lascaux, Altamira and a number of the African sites, the idea of composition appears to be absent. But when confronted with prehistoric art, we must always remember that we see it through eyes schooled in twentieth-century preconceptions. What you and I see on those walls is not what the artists and their fellows saw, for art loses much of its meaning in the absence of social context.

Most cave art depicts animals, but there are also some abstract motifs: groups of dots, meandering lines, rectangular grids in which the individual squares are sometimes filled in with different colours. By far the most common animal depicted is the horse, followed by bison and oxen: together these form almost sixty per cent of all animal images. The rest are mostly deer, mammoth and ibex, with reindeer, certain antelope, goat, boar, rhinoceros and a number of carnivores, such as lion, hyaena, fox and wolf. There are very few birds or fish, although we know that these creatures were an important part of the life of Ice Age people from their food debris. Most

The major sites of finds of Ice Age art in France and northern Spain.

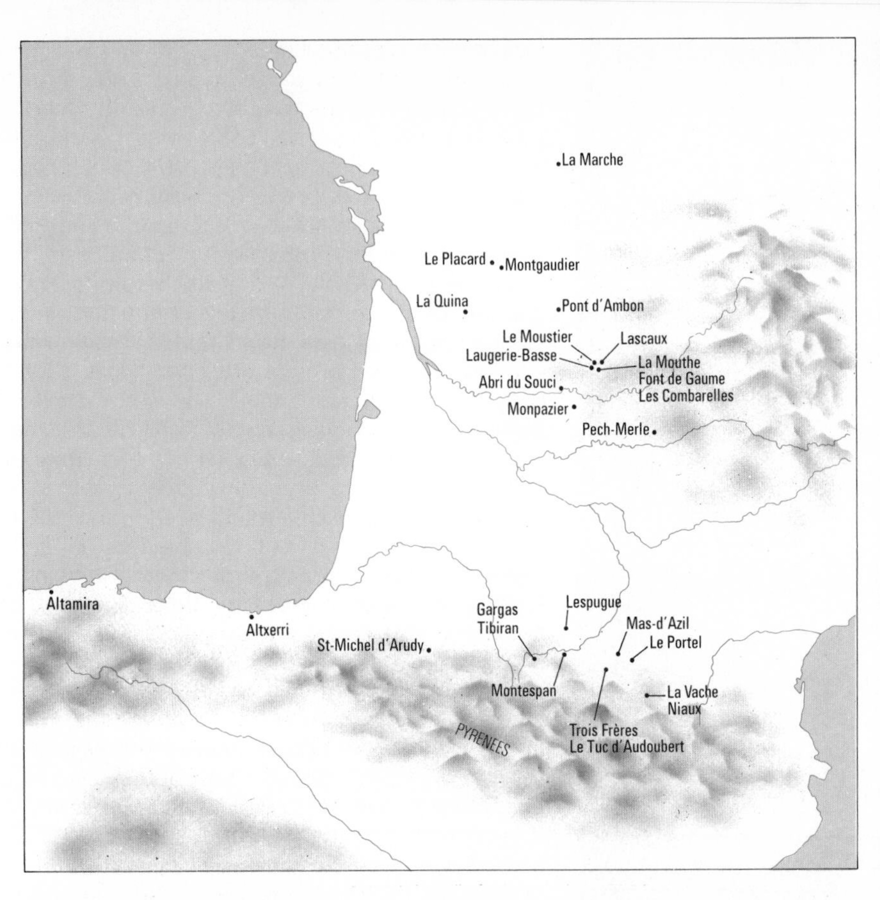

curious of all is that there are very few human representations. Those that do exist are almost all sketchy lines suggesting the human form, rather than naturalistic portraits such as one knows the artists were capable of. Only at the remarkable cave of La Marche in western France are there images that even approach portrait sketches, and many of these are rather more like caricatures. There are, however, prints in paint of human hands and, in many caves, the outlines of hands, made by pressing the hand on the cave wall and applying paint around it. These seem to say 'I was here'.

Although most popular attention goes to the paintings the small portable objects, such as engraved bones, antlers and stones, were also a vital aspect of Ice Age art. Thousands of beautifully carved and engraved objects have been discovered over the last century, many of them coming from the decorated caves. Academic neglect over the years, which a number of scholars are now rectifying, means that little can be said about this 'portable art'. A number of things are clear, however. For instance an engraved bone can be personal art, worn, perhaps, as simple adornment, or as a badge of belonging to a particular band. There are many other possibilities too, some of which I will touch on later. But with portable objects of this sort, the art can go to the people, rather than the people having to go to the art as is necessary with the wall pictures.

A striking difference between the two forms of art is that there are very few depictions of animals on portable objects. Most portable art carries some kind of geometric pattern: arcs, sets of dots, chevrons, zig-zags and so on. In the wall art of Altamira there is a marked emphasis on bison, whereas there is not a single bison image on portable art from that cave.

Instead, deer are popular engraved figures here, and the cave-floor debris shows that these animals were important as a source of food to the people who gathered at Altamira. Another interesting point is that the portable art from different parts of Europe does not show the sort of uniformity exhibited by the cave paintings, perhaps reflecting its more personal nature.

One of the great puzzles of Palaeolithic art is the scant treatment that reindeer receive in the cave paintings. From the litter of bones in the decorated caves, it is clear that reindeer were prominent in the diet: sometimes reindeer bones make up ninety-eight per cent of the animal remains at living sites. And yet paintings of these animals are extremely rare. Why is this? Claude Lévi-Strauss once observed of art among the San and the Australian Aborigines that certain animals were most frequently depicted not because they were 'good to eat' but instead because they were 'good to think'. It is hard for us to penetrate the world of the cave artists. Our perspectives are not theirs, and our beliefs and conceptions of the world are certainly different. But whatever it was that motivated the strokes of their brushes and the incisions of their engraving tools, there was an astonishing durability about it.

I have mentioned that there was a degree of similarity in the cave paintings throughout Europe. There also appear to be common threads running through time. Although one can see an element of development in the art through its 25,000-year period, the changes were not dramatic. Overall, there is more sense of stability than of change, and this contrasts sharply with the relatively rapid changes in style of tool industries during this period. The great French prehistorian and expert on cave art, André Leroi-Gourhan, has said: 'In the life of a society models of weapons change very often, models of tools change less often and social institutions very seldom, while religious institutions continue unchanged for millennia.'

The meaning of Ice Age art

As soon as the cave paintings were accepted as authentic products of Ice Age people, prehistorians began to try to understand the meaning of it all. The year after Emile Cartailhac published his confession that he had been wrong about Altamira, he invited a young French priest, Abbé Henri Breuil, to visit the cave with him. The Abbé Breuil was, at the age of twenty-six, already an established expert on the terminal part of the Ice Age, and he was destined to become one of France's greatest prehistorians. He spent many years making exquisite copies of cave art throughout Europe, and he began an analysis of the stages of its development. This latter project was made extremely difficult because of the absence of reliable dates for many of the decorated caves. This problem persists, even today, when far more sophisticated dating techniques are available.

The Abbé Breuil interpreted Ice Age art as a medium of hunting magic, a supernatural means to ensure 'that the game should be plentiful, that it should increase and that sufficient should be killed. . . .' The disorganized jumble of images, Abbé Breuil thought, implied that they had been painted as part of a series of ceremonies performed in preparation for the hunt. The geometric patterns were, according to the priest, diagrammatic depictions of traps and snares. He believed that most of the decorated caves were sacred localities, set aside for ritual occasions. Like most prehistorians of the time, he envisaged life in the Ice Age as a perpetual struggle to obtain sufficient food, hence the need to call on magical powers to help with the hunt. The Abbé Breuil proposed that the art faded into oblivion 10,000 years ago, because the teeming herds dispersed as the ice sheets retreated. The era of the Ice Age hunters passed, and with them went their magic.

The Abbé Breuil believed that the decorated caves were sacred places where the hunters came to perform a ceremony before any major hunting expedition: each time they would add new animal figures, resulting in the chaotic jumble of images seen in many caves. On this wall in the Niaux cave system, bison and horses are mingled, and the artists seem to have added arrows piercing their flanks. The Abbé Breuil's 'hunting magic' theory was an attractive explanation for the painted caves, but there were many difficult questions which it left unanswered. There was a particular problem regarding reindeer: the food debris of living sites show that they were an important prey animal, but they are very poorly represented in cave art.

This view of Palaeolithic art persisted until just a few decades ago. There are distinct problems with it, not least of which is why the reindeer, which we know was the most important of the hunters' prey, should be so poorly represented in the cave paintings. And if Ice Age art was simply hunting magic, how does one explain the fact that only about ten per cent of the animals are shown as having been killed?

The discovery, in 1940, of Lascaux, with its relatively coherent friezes, was something of a blow to the Abbé Breuil's proposal, which was based in part on the apparently chaotic jumble of painted figures in cave art. Were the orderly and composed paintings of Lascaux and of Altamira something out of the ordinary, or did they just state clearly what was less apparent, but nevertheless present, in other decorated caves? André Leroi-Gourhan, and, independently, Annette Laming-Emperaire, considered that there was indeed a structure to the painted caves, if one knew how to look for it. 'Indeed,' said André Leroi-Gourhan, 'consistency is one of the first facts that strikes the student of Palaeolithic art. In painting, engraving and sculpture, on rock walls or in ivory, reindeer antler, bone and stone, and in the most diverse styles, Palaeolithic artists repeatedly depict the same inventory of animals in comparable attitudes. Once this unity is recognized, it only remains for the student to seek ways of arranging the art's temporal and spatial subdivisions in a systematic manner.'

The many images of horse and bison in Ice Age art are there for all to see, but according to André Leroi-Gourhan and Annette Laming-Emperaire, this does not simply reflect an artistic obsession with these animals. The two prehistorians see a deeper significance and one that encapsulates a model of society. The 'model' is the duality between male and female, and the argument suggests that symbolically male and female images are distributed separately within each cave, thus reflecting a fundamental division in the world. To André Leroi-Gourhan the bison represented the female element while the horse represented maleness. Annette Laming-Emperaire, incidentally, saw the attributions the opposite way around. Also female in André Leroi-Gourhan's scheme were the ox, the mammoth and certain geometric signs which were approximately in the form of female genitals. Stag, ibex and phallus-like signs were taken to be male symbols.

In his extensive and meticulous researches André Leroi-Gourhan studied sixty caves and more than 2,000 images. Female symbols, he concluded, were predominantly in central parts of caves whereas the more peripheral zones were mainly occupied by male images. In other words, cave art should be viewed not as a random collection of paintings of animals, but as an ordered symbolic projection of the hunters' world. The sexual duality was the most obvious facet to strike André Leroi-Gourhan's eye, but he allows that the division may represent some other aspect of Palaeolithic life of which we are completely unaware.

There are many criticisms of this interpretation of Palaeolithic art, not least of which is the question of why a male bison should be thought of as representing 'femaleness' and a pregnant mare as 'maleness'. And in his analyses, André Leroi-Gourhan concentrated on the types of images and was not influenced by the *number* of, say, bison pictures in a particular section of a cave. Nor did he take into account the size or colour of the animals. These properties may have been important to the artists when they did them, or they may not have been. But André Leroi-Gourhan's assumption that the type of animal, rather than its individual characteristics, was most important may have allowed him to see more order in the caves than in fact exists.

However, the emphasis on *context* as well as *content* in the study of Ice Age art that André Leroi-Gourhan and Annette Laming-Emperaire introduced in the 1960s must be seen as an important advance in our perception

A horse from Lascaux cave. The frequency with which horses and bison appear in the cave paintings led to the idea that these animals were of symbolic significance. Independently of each other, Annette Laming-Emperaire and André Leroi-Gourhan, came to the conclusion that they symbolized the division of society into male and female elements. According to this theory, the arrangement of the images within a cave represented the structure of the hunters' society.

of life in the past. It set the scene for much of the work that followed. For instance, Margaret Conkey has recently speculated that the painted ceiling at Altamira may reflect the social organization of the people who went there.

The core of the composition at Altamira is a group of bison whose physical attitude has been interpreted by some experts as representing charging animals. Others, however, suggest that the bison are depicted in the throes of death or in the process of giving birth. The bison are surrounded by a small number of other creatures, such as wild boars, a horse and a female red deer. These images, Margaret Conkey and others suggest, may have nothing explicitly to do with hunting, but instead may represent the people within the social group. If one views the central bison as female animals giving birth then these could symbolize the central position of women in hunter-gatherer society. The men are somewhat peripheral to this society, inasmuch as they form groups and leave the home base for days at a time while hunting. Could the horse, the deer and the boars represent men who, while being part of the hunter-gatherer band, are in a sense peripheral to it, Margaret Conkey wonders? Many anthropologists see this 'matrifocal' arrangement as being a common theme in foraging societies.

The notion of art reflecting society in this way is certainly plausible, and Margaret Conkey sees it as extending to other levels in society too. She has been analysing in fine detail the design elements used in the engravings and carvings on portable art from many sites in northern Spain. Although to the untutored eye the portable art of this region looks, if not uniform, then at least very similar, she has been able to detect features that distinguish one location from another. At Altamira something interesting happens because there is a coming together of design features from many nearby localities. The range of stone tools found at the cave are also a collection of styles from outlying areas. Altamira, it seems, was a major site where people from many different places congregated at certain times of the year, most probably in the autumn when red deer and limpets were plentiful. People who were, for most of the year, dispersed among the hills and valleys of Cantabria came together seasonally, resulting in a pooling of technology and of art. The reason for the annual aggregation may have been to exploit rich and concentrated food resources. Or there may have been social reasons: perhaps they sought an opportunity to mix with a larger group of people for everything from exchanging gossip to seeking a marriage partner, just as hunter-gatherers do today.

What does this have to do with the ceiling at Altamira? The images could, according to Margaret Conkey, represent the aggregation of different social units which, though distinct, have a common purpose. Chicago anthropologist Leslie Freeman is sympathetic to this idea and proposes that Altamira and other large decorated caves 'might have served as periodic centres of assembly where seasonal ceremonies were conducted on behalf of the congregated population of a large surrounding area'. The Altamira ceiling may therefore have a double meaning, representing social relations *within* bands and *between* them. It may, of course, be nothing of the sort, but the proposal is certainly worth considering.

Ancient rituals in the caves?

Intimations of ancient ritual abound in the decorated caves of Europe, but there is probably none more dramatic than the scene deep within a tortuous cave system under the foothills of the Pyrenees, near Ariège. Here, the little River Volp runs out of the tree-shaded entrance of a cave known as Le Tuc d'Audoubert which is connected with a second cavern named Les Trois Frères. Almost seventy years ago the three sons of the late Count

Henri Begouën set out to explore Le Tuc d'Audoubert, entering the cave on a simple raft. Their journey was full of adventure and ended in a fantastic discovery.

Lighting their way with small hand-held lamps, the boys soon saw that they should pull up their raft at a small gravel beach and proceed on foot. Passing 20 metres (65 feet) along a passageway, they found themselves in a large and beautiful chamber, filled with the most fantastic stalactites. The Begouën boys then scrambled 15 metres (50 feet) up a steep tunnel that led off from the chamber, and continued their explorations through numerous galleries, chambers and low tunnels, wriggling and feeling their way into the depths of the earth. The boys had noticed bones of cave bears and other animals littering their path, as well as scatterings of prehistoric stone tools. But in all their wildest dreams the boys could never have anticipated what they were soon to see in the flicker of their lanterns: two moulded clay bison each almost a metre long, propped up in the middle of a low, round chamber. This ancient sculpture had been there, untouched and unseen for about 15,000 years.

I had seen pictures of these famous bison before I visited the cave, but nothing could have possibly prepared me for the feeling of incredible awe that I had when I first saw the masterpieces deep in that French hillside. When the Begouëns had first entered the chamber in October 1912, a third, smaller, bison lay on the floor close to the main figures. This was later removed and placed in a Paris museum for safekeeping, although a cast of the original stands in its place in the cave. The outline of another minute bison was sketched into the floor. What had happened in that chamber those many thousands of years ago? Surely Palaeolithic people did not make this very difficult journey into the bowels of the earth simply to exercise their artistic skills? The journey must have been far more hazardous than it is today since they can have had nothing to light their way other than animal-fat candles which, if extinguished, could not be relighted inside the cave.

When Count Begouën, an archaeologist at the University of Toulouse, examined this remarkable cave with some of his colleagues he found other evidence to suggest that the cave had been a scene of ritual. Nearby, in a side chamber some 25 metres (80 feet) from the bison, there is a pit from which the clay was taken to make the figures. The people who sculpted the bison left their footprints in the wet clay of the side chamber but what is extraordinary is that the prints are only of the heel of the foot. For some reason, the artists took great care not to leave the marks of their toes. In one corner of the chamber lie five 'sausages' of clay. Were they to be used in moulding clay figures perhaps, or were they symbolic phalluses? No one will ever know.

Another cave in the Pyrenees, the Grotte de Montespan, also contains clay figures, although here the workmanship is not as high as at Le Tuc, and the figures have deteriorated more with age. Fragments of a clay lion and clay bears lie against the walls of a cavern some 2 kilometres (1·2 miles) from the cave entrance. Once again there is a scattering of footprints in the clay floor, some of which are those of children.

Is it significant that in both these cases a river runs through the cave? British archaeologist Paul Bahn thinks it might be. Rivers, pools and lakes, he points out, are often endowed with strong spiritual properties by technologically simple people throughout the world. And the combination with caves, representing an entrance to another world, imbues the water with even greater powers. 'There is every reason to assume that water played a major role in the Upper Palaeolithic system of beliefs, and would therefore

The clay bison from Le Tuc d'Audoubert. Even with the aid of electric lamps and an inflatable raft, the journey into the cave is difficult. Strong motives must have impelled the artists to penetrate the cave with nothing more than simple oil lamps or wax tapers to light their way.

be a factor incorporated in any ritual art of that period,' Paul Bahn proposes.

Returning again to Le Tuc, one finds more echoes of a mystical past in a second cavern, Les Trois Frères, named after the Count's sons. The three brothers once again penetrated the hillside's hidden chambers two years after they made their remarkable discovery at Le Tuc. This time, after squirming their way through tiny passageways they found a cavern whose walls were festooned with delicate animal engravings. The Palaeolithic artists must have spent many patient hours etching the images on these walls, their simple wick lamps throwing flickering shadows in the narrow passages.

Les Trois Frères is one of the most intensely decorated prehistoric caves, but perhaps its major interest, and certainly its most startling image, is that of a chimera of human and animal features, the so-called 'sorcerer'. In this painted and engraved image a pair of antlers perch above a bearded face

from which two owlish eyes stare fixedly. The creature has a horse-like body and tail, with bear's front paws and distinctly human-like hindlegs. Human male genitals hang oddly out of position beneath the tail. If the creature itself is odd, so too is its position for it stands virtually on all fours and is situated on the wall of a 'chimney' suspended above the richly decorated cavern.

Is this a picture of a Palaeolithic sorcerer dressed in animal skins and deer's antlers, as Abbé Breuil and Count Begouën initially believed? Or is it part of the artists' mystical universe, an expression of the duality of the animal and human worlds? Creatures which show a mixture of human and animal characteristics, or are an amalgamation of two different animals are a common theme in many different religions. Feathered serpents and owl-monkeys were among the supernatural beings of many South American cultures for example. The sorcerer of Les Trois Frères and the sorcerer of Lascaux may well be ambiguous creations of this sort but Margaret Conkey suggests another interpretation of the Lascaux and Trois Frères images. They might, she speculates, be meant to emphasize not the separateness of the human and animal worlds, but rather their continuity. If Palaeolithic people conceived of themselves as very much a part of the world around them, then this interpretation is clearly plausible. There must be many possibilities, and no one can be sure which is correct.

The sight of the long-preserved footprints of young children in barely accessible nooks of extensive cave galleries is particularly intriguing. What could young children have been doing deep inside these dangerous caves? A recently discovered section of the vast cave system at Niaux in the Ariège Valley gives another tantalizing glimpse of this puzzling aspect of ancient life. At one point three children crouched low in a small side chamber, barely a metre (3 feet) high. Their prints remained crisp and undisturbed there for at least 10,000 years. At the nearby cave of Fontanet a group of children appears deliberately to have left their foot and hand prints in a gallery leading to a large chamber. The chamber, incidentally, contains food debris, which is rare inside decorated caves other than simple rock shelters. Perhaps some feast accompanied the ceremonies that were performed in front of the painted and engraved images of bison and ibex. This chamber has another unusual feature: in addition to the animals, humans with great bulbous noses are depicted on its walls.

Strangest of all, however, are prints of hands, mostly shown in outline by paint being sprayed against the hand held to the surface of the wall. Presumably the paint was 'sprayed' by taking it into the mouth and then blowing it out through a fine straw made from a plant stem. Such prints are to be found in more than twenty caves throughout France and Spain. Mostly such images are few in number and are relatively inconspicuous. In three caves, however, the practice of hand printing has a strange and disturbing twist to it: at Maltravieso in northern Spain and at Tibiran and Gargas in the Pyrenees, many of the prints reveal some degree of mutilation of the fingers. Gargas is the most dramatic case, where clouds of red and black stencilled prints show the most astonishing deformities. More than half of them have lost the upper joints of all four fingers, while others show partial or total loss of one or more joints. Strangely, the thumbs appear to have escaped injury. There are also several prints of young children, whose intact hands were held up to the wall by an adult.

There is no doubt that the hands of the people at Gargas were mutilated: the images were not a trick of painting or spraying technique. The question is, what caused it? Was the community ravaged by disease, was frostbite responsible, or did the people of Gargas practise a bizarre ritual that involved

The 'sorcerer' from Les Trois Frères cave. The Abbé Breuil and Count Begouën interpreted the figure as a man dressed in animal skins, a sorcerer engaged in the hunting magic ceremonies. Current interpretations see this strange figure as representing the relationship between the human and the animal worlds.

mutilation? There is simply not enough evidence to answer this question, and the mystery will probably remain forever unsolved.

New interpretations

During the mid-1960s, Alexander Marshack, a journalist turned archaeologist, began applying techniques to Palaeolithic paintings, carvings and engravings that no one had developed fully before. He photographed the paintings using infrared and ultraviolet light, and he examined the lines and cuts of engraved bones and stones under a microscope, looking for minute clues about the way the objects had been worked. Alexander Marshack caused more than a little stir among academic archaeologists when he suggested that a series of tiny engraved pits on a 30,000-year-old bone implement indicated that its Ice Age owner had been noting the changes in the lunar cycle. The snake-like procession of pits had previously been taken as nothing more than 'doodling'. But through the microscope Alexander Marshack had seen that as many as twenty-four different tools had been used to nick out the sixty-nine pits. Clearly, the snake of pits was the result of a series of acts spread over a period of time. Alexander Marshack also worked out that the arrangement of the curves in the engraved 'snake' coincided with the changing phases of the moon.

The idea remains an audacious challenge to the academic community. Meanwhile, Alexander Marshack has moved on to develop a consuming interest in Palaeolithic symbolism, both in the naturalistic paintings and engravings and in the geometric motifs. 'I'm concerned with the way that art is *used* in society,' he says. 'Art isn't simply a product of people's culture. It is a mark of that culture. And one of the fascinating aspects about human history is the way these marks of culture evolved and became intertwined with more basic subsistence activities, such as obtaining food. Art goes farther than that though. It involves cognitive skills that far outreach anything that's needed *simply* for subsistence. A Mozart, a Michaelangelo, the building of the pyramids or the Parthenon – all are intellectually far in excess of the subsistence strategies used in those cultures.'

One artifact that Alexander Marshack has studied is a small bone knife from a cave known as La Vache, that apparently had never been used for cutting anything. The reason, he suggests, is that it was meant for 'higher' purposes. Engraved on one face of the blade is a doe, some wavy lines suggestive of water, three flowers in bloom, and an ibex head with one horn crossed out. On the other side of the blade is the head of a bison bull bellowing as in the autumn rut, some leafless branches, some nuts, and a dying flower. To Alexander Marshack, the knife is a marker of the seasons, autumn on one side of the blade and spring on the other: 'Perhaps the cross on the ibex horn represents a *symbolic* killing. I would see objects such as this being used in rituals that signal the arrival of certain seasons.'

An even more striking seasonal composition, in Alexander Marshack's eyes at least, is shown on a section of reindeer antler found at the Montgaudier cave in southwest France. Since its discovery in the 1880s the images of various creatures on it had been taken to represent hunting magic. But Alexander Marshack's detailed observations revealed other possibilities. The images of a bull seal and a smaller female seal appear in amazing detail. Next to the seals there is a salmon whose lower jaw has a hook of the sort displayed by the males in the migratory run upstream to

A handprint from Pech-Merle. Here the hand is intact, but in certain caves the handprints reveal severe mutilation of the fingers that may have been the result of frostbite in the bitter Ice Age conditions.

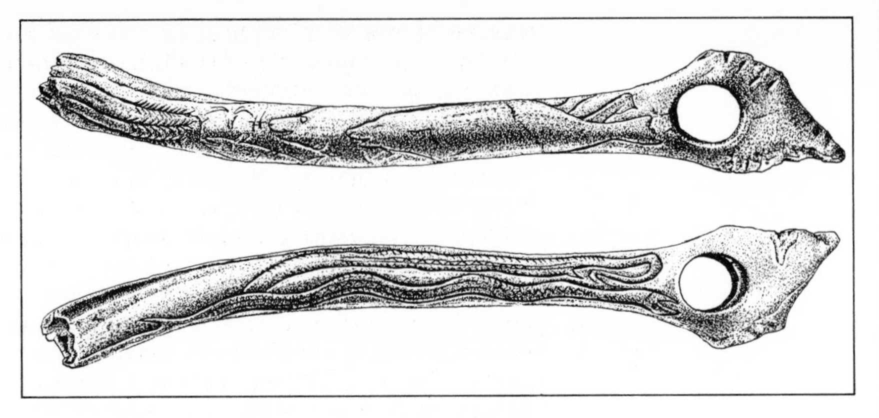

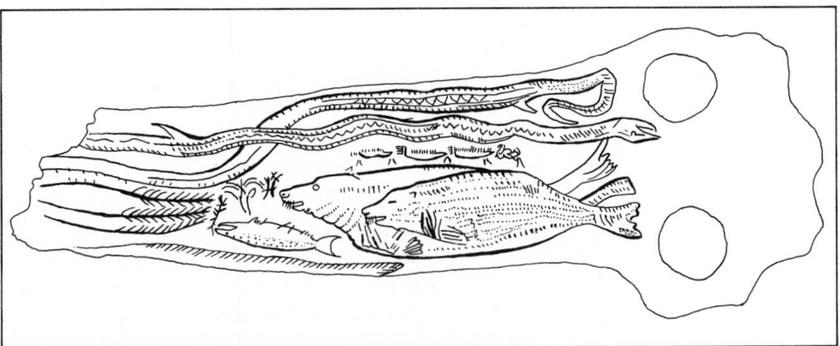

A baton, carved from reindeer antler, which was found in Montgaudier cave in southwest France. The two drawings at the top show the baton from different sides. The engraved images continue all the way round, and the diagram below shows the baton opened out.

mate. To the left of the salmon are three lines that archaeologists had assumed were harpoons but which Alexander Marshack's examinations show to be stems with leaves: the barbs run in the wrong direction for them to be harpoons. A tiny flower is in bloom above the female seal. Next to the bull seal are three enigmatic objects said to be many-legged water creatures. Two snakes, displaying genitalia lie almost coiled together. Last, again seen clearly only under the microscope, a schematic ibex head viewed from the front has a cross on its forehead. The engraving is exquisite and it rolls right around the curved surface of the antler.

'I see the Montgaudier baton as a clear seasonal composition,' says Alexander Marshack. 'The seals are ready for mating, and so too are the snakes and the salmon, and the flowers are in fresh bloom. It's interesting to compare the very realistic engraving of most of the images with the highly schematic ibex which, I suggest, has been symbolically killed with the cross. The engraving suggests to me an act of killing, not for food but as a symbolic ritual related to the coming of spring.'

I talked in Chapter 8 of the signs of prolonged use of a small ivory horse statuette from Vogelherd in Germany. The smoothly polished features of the 32,000-year-old figure indicate a history of use, either as a personal talisman or more formally in some social ceremony. An angle is cut into the shoulder of the horse, and Alexander Marshack sees this as representing a symbolic killing. It was with the Vogelherd horse, and other statuettes from the site, that he first developed the notion of the use and re-use of art. He also sees it in wall paintings such as the curious images in the cave of Pech-Merle, which show two large horses bespattered with red and black dots. With the aid of infrared photography he was able to tell that the outline of one horse was first painted and then sequentially scattered with groups of dots, followed by a similar sequence for the second horse. 'Like the

Vogelherd horse that had been used over a period of time and then symbolically killed, this wall at Pech-Merle had been used in a deliberate and repeated sequence,' Alexander Marshack claims. 'I don't know what was going on, but those horses certainly weren't meant to be killed and eaten. Something more was happening. Something we don't yet understand.' Examples of reworking of painted and engraved images are common in Palaeolithic art: heads may be redrafted, legs redrawn in slightly different positions or the whole outline changed. Alexander Marshack's interpretation is that the art object is almost 'alive' and is part of a ceremony. There are, however, alternative explanations. For instance Michel Lorblanchet, one of France's leading experts on Palaeolithic art, speculates that such 'retouching' may simply have been part of artistic convention. The sketching of heads and legs in many different positions may, he suggests, be meant to convey a sense of movement, a feeling of activity in an otherwise static image. At first sight this might strike one as simply an amusing idea but only because it does not fit in with our conventional ideas about art. As I said earlier, the meaning of art depends to a great extent on its social context.

The human form

Looking at the span of Palaeolithic art as a whole, one sees some strong images: the bulls of Lascaux, the bison of Altamira, the horses of Pech-Merle. In all these, the colours are rich, the lines are bold and confident, and the eye for detail is acute. Against such a rich representation of the animal world of the Ice Age, the human image is conspicuous by its scant and impoverished treatment. Nowhere is there a portrait of a human face equivalent in detail to one of the giant bulls at Lascaux. It is almost as if there was a taboo against the representation of the human form. Perhaps there was such a cultural convention. If so, it cannot have been complete because at La Marche, in western France, portraits of a kind do exist, but they are rudimentary compared with the animal pictures.

The horses of Pech-Merle. With the aid of infrared photography, Alexander Marshack concluded that the outline of one horse was painted and then covered with groups of dots, followed by a similar sequence for the second horse. He believes that the horses were, to the cave artists, almost 'alive' and that a ceremony was acted out around them.

La Marche is a south-facing rock shelter poised on a slope above a small stream. Twelve thousand years ago alder and lime grew close by the stream. In other parts of the valley copses of oak, box and ash were fringed with bracken. Pansies, bellflowers, garlic and oregano flowered in spring and summer, while redcurrants and hazelnuts were to be picked later in the year. It was indeed a pleasant site, and an extraordinary one too. For some unexplained reason the people of La Marche, who were extremely prolific engravers, featured human subjects in their work more than animals – quite the reverse of the typical pattern.

In all there are fifty-seven engravings of isolated human heads, and a further fifty-one less complete heads and bodies. This one site accounts for more than a quarter of all human representations so far known from Ice Age art in Europe. The heads are profiles, some with unbelievably large jaws and comically turned-up noses: presumably these are caricatures. Many, though, project the impression of personal portraits and they are usually identifiable as males or females. The hair is mostly short, and in a few cases it appears to be finely plaited. Ten men wear beards, and there are three certain moustaches. One individual has a headband and a dozen hats are to be seen. La Marche certainly gives us the clearest picture we have of the Ice Age artists themselves.

The best-known representations of the human form from prehistory, however, are the so-called Venuses, statuettes with bulbous buttocks and breasts which supposedly embody a fertility or mother-god image. Statuettes of this type are certainly very striking in their emphatic sexuality. The sense of super-femaleness in the exaggerated curves of the little ivory Venus of Lespugue, France, is unmistakable. And the 5-centimetre (2-inch) tall Venus from Monpazier in the Dordogne, in addition to the protruding buttocks and breasts, has a distinct vulva boldly carved into the figure. But it turns out that statuettes displaying exaggerated female features are in a tiny minority. Of the many hundreds of carved figures so far discovered throughout Europe, some can be identified as female, although most of these have natural rather than exaggerated proportions, some are clearly male, but most are, to our eyes at least, sexless. The idea of a continent-wide cult of the mother-god, symbolized by the bulbous 'Venuses', appears to have been greatly overstated.

This new conclusion is due mainly to the work of British prehistorians Peter Ucko and Andrée Rosenfeld and to the notable French researcher Léon Pales. These workers analysed in greater detail than ever before the full range of statuettes found. Commenting on his observations, Peter Ucko said: 'The figures represent a heterogeneous group of individuals rather than a single individual, human or deity.'

Many of the figures have the impression of being 'rough sketches' of the human form, rather than carefully sculpted representations. In the main, there is no sense of portraiture such as is seen in the La Marche engravings. Even in the more skilfully worked pieces facial features are usually absent. This may be because there was a taboo against capturing the true image, and therefore the spirit or soul, of an individual. But why should legs and feet be so sadly neglected? Virtually all prehistoric statuettes have truncated stumps for legs. This contrasts markedly with the great attention paid to the legs and feet of animals in the cave paintings and engravings.

In eastern Europe and Russia, where the sparsity of suitable cave shelters probably contributed to the virtual absence of wall art, one sees a greater emphasis on the creation of figurines. A particularly fine collection of statuettes was found among the remains of six huts at Maltà, on the terraces of the River Belaya in eastern Siberia. The people of Maltà included an

The Venus of Lespugue. The original statuette was carved from ivory and, when found, it was broken and stained with age. This cast of the figure reproduces the beautiful symmetry of the original curves. Although the Venuses, with their exaggerated sexual features, are the best known of the human representations of Ice Age art, they are not, in fact, all that common.

The human form is rarely depicted in Ice Age art but, where it does appear, the figures are sketchy and roughly drawn. This contrasts markedly with the skill and artistry shown in the paintings of animals.

extraordinary amount of detail in their work, and in one case there is a strong impression of that elusive 'portrait': the tiny ivory figure shows eyes, nose and mouth in unusual clarity.

The principal features of Ice Age art in eastern Europe and Russia, however, were geometric patterns. Alexander Marshack explains the different artistic expressions in the following way: 'Ice Age art was very much the result of what was regionally possible. In southwest France and northern Spain the extensive systems of limestone caves permitted the creation of painted wall art. And as the people there shared a common ecology of interconnected river-valley systems inhabited by a similar range of animals, it's natural that their art should reflect this. Their symbol systems were expressed in naturalistic images whereas in eastern Europe and Russia people developed different expressions for their symbol systems. These systems were mainly in the form of geometric and schematic patterns. It doesn't mean that the two forms of expression were different in cognitive complexity. Which is more complex, a computer or a Mozart symphony? You can't say. They are different manifestations of a certain intellect. The important fact about Ice Age art was that people were developing symbolic expression of the feelings, beliefs and social systems that were important to their lives.'

The end of an era

The Age of Art lasted for 25,000 years and it ended 10,000 years ago. Before the Age of Art we find signs of some sort of artistic expression, although much less well developed, stretching back for many thousands of years. But when the Ice Age ended, Ice Age art with its vibrant and dynamic images of animals disappeared almost without issue. Why did this flourishing tradition of artistic expression end so abruptly?

To answer this question one would have to be a great deal more certain about the motivations of the artists when they were at work. One simple explanation is that the warmer climate meant that people no longer used caves. But this is just too simple: they could have entered the cool caves for rituals and social ceremonies if they had wished, and all the evidence suggests that the cave paintings did have ritual significance.

The more clement conditions saw the disappearance of the vast herds of cold-adapted animals, such as the all-important bison. Woodland began to cover hills and plateaus that previously had been open grassland. The life of the hunter-gatherer bands in Europe must have changed dramatically over a period of just a few millennia. And with the final arrival of agriculture, social structures altered even more.

Alexander Marshack suggests that extensive contact between groups of people with similar ways of life was essential for the vitality of the Age of Art. The loss of the social network stretching over very broad geographical regions may have sapped that vitality and led to the abandonment of cave painting. There must, however, have been many important threads running through the lives of these people, social, ritualistic and aesthetic, that would still have found expression in their art. The temptation to seek an all-embracing solution to the meaning of Ice Age art and the reason for its disappearance should be avoided.

Art itself did not vanish with the termination of the Ice Age, but it changed, both in content and in spirit. Geometric art flourished and evolved in many areas. In those regions where representational images reappeared, there was a distinct shift in focus from the animal to the human world, but there was no exaltation of the human form analogous to the animal imagery of the Ice Age. Rather, there emerged an obsession with depicting confrontation between people. Scenes of battles became common, representing perhaps a record of a past conflict or an attempt to enlist supernatural aid in a forthcoming fight. A new age was dawning, and art may have been expressing its hopes and anxieties.

11 Hunting in Transition

In February 1980 Paul Bahn, an archaeologist from the University of Liverpool, set out for Paris with a rather unusual task in mind. He had arranged to visit a number of museums in Paris that have rich stores of fossil bones from Ice Age Europe. Paul Bahn is particularly interested in the lifestyle of the cave dwellers of 40,000 to 10,000 years ago, the earliest modern men, whose art we looked at in the previous chapter. Somewhat surprisingly, he was going to Paris to look at the museums' collections of prehistoric horse teeth. By chance he came across exactly what he was looking for: the complete front section of a lower jaw in which the teeth show a characteristic pattern of wear. The British archaeologist's discovery in the Institut de Paléontologie Humaine reopens a controversy that raged bitterly in France at the turn of the century, and it may force many contemporary prehistorians to rethink their ideas about life in the Ice Age. These teeth suggest that as long as 30,000 years ago our ancestors were tethering horses.

The traditional view of human prehistory is that there was a long period when our ancestors lived very successfully by hunting-and-gathering, and then, about 10,000 years ago, agriculture was introduced. This traditional theory maintains that agriculture was invented independently in many different parts of the globe and spread very rapidly from those places through most human populations. The transition is often called the Agricultural Revolution.

The change was indeed revolutionary, since the two styles of life produce very different social orders and economic structures. Compared with the pace of change among human populations prior to 10,000 years ago, the adoption of agriculture, as revealed in the archaeological record, does seem to have been virtually instantaneous. But the Agricultural Revolution may not have been such a dramatic switch in lifestyles as has generally been thought. There is a very real possibility that the people of the Ice Age exerted more control over their food resources than is normally implied by a hunting-and-gathering way of life and husbanded their food resources to a much greater extent than has been realized. Thus, the Agricultural Revolution may only have involved a greater commitment to a mode of production with which people were already familiar rather than the introduction of something totally novel.

A Lapp sleigh drawn by a reindeer. The Lapps and certain Siberian tribes follow the reindeer herds as they migrate. They hunt the reindeer for meat, but also domesticate some animals for milk and transport. Some prehistorians now believe that a similar way of life, intermediate between hunting and true pastoralism, may have begun to emerge during the last Ice Age.

Advances in technology

Throughout human prehistory, tools tell us most about the activities of our ancestors. For most of the past, however, the outstanding characteristic of man's tools was the lack of change over long periods of time. Only with the emergence of fully modern humans during the last Ice Age did the pace of change begin to increase, with new 'cultures' appearing roughly every

5,000 years, rather than every 500,000 years as was common in the earlier stages of human evolution.

The technological change has three aspects: the form of the implements, the material from which they are manufactured, and the number of tool types produced. Many of the tools in the later cultures are finer and more delicately made. They are clearly intended for more intricate work than those in previous ages. Tools such as 'awls', 'projectile points', 'hide-burnishers' and 'needles' make an appearance. Delicate implements were manufactured regularly by people in earlier periods, but they were not as significant a part of the typical toolkit. Many of the new types of implements, particularly sharp points and tiny chisel edges, are made from bone rather than stone, the advantage of bone being that one can craft slender but strong implements from it. Indeed, in many places stone tools become only a minor component of the toolkit. As for the number of separately identifiable objects, the cultures during the later part of the Ice Age encompass perhaps a hundred different items compared with around sixty for cultures prior to 40,000 years ago.

But the crucial element of the toolkit is not the number of items it contains, nor is it the shape of the implements: what matters most is how the tools are employed. A hint of the great intelligence that accompanied the more refined toolkits of 40,000 years onwards comes from some extremely important analytical work on two South African coastal cave sites carried out by Chicago anthropologist Richard Klein. The people who inhabited the Klasies River Mouth cave, situated 450 kilometres (280 miles) east of Cape Town, exploited their rich environment with a considerable degree of success when they lived there between 100,000 and 70,000 years ago. They hunted a number of large animals, including Cape buffalo, giant

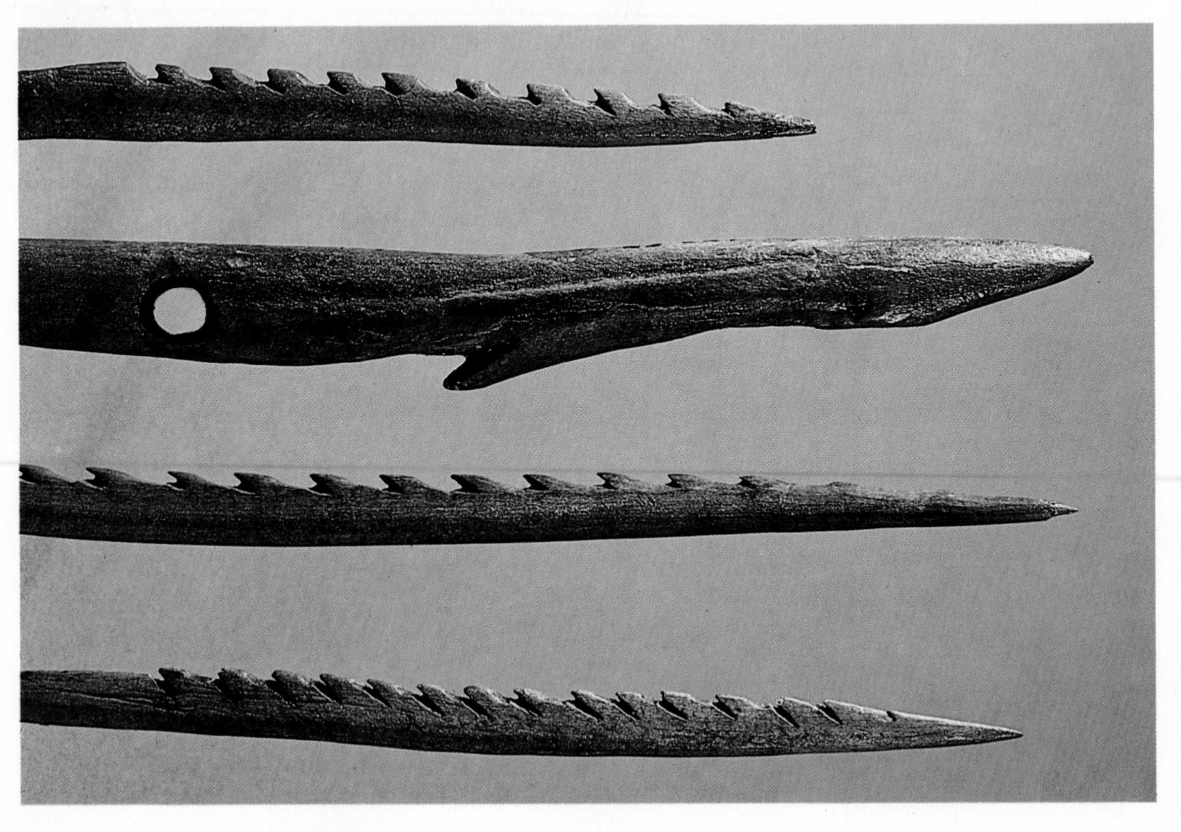

With the emergence of fully modern man some 40,000 years ago, technology advanced significantly. Precise and delicate tools appeared, including needles, and more tools were made from bone than ever before. Bone is less brittle than stone and can be used to make implements that are slender and finely pointed but also strong. These harpoons were made from antler, a raw material with the same sort of advantages as bone.

buffalo, eland, blue antelope and black wildebeest. They regularly collected shellfish from the nearby rocky shore, and Richard Klein points out that 'this is the earliest evidence found anywhere in the world for systematic use of aquatic resources'. Although they included sea birds on their menu, this was a relatively rare occurrence.

One hundred kilometres (60 miles) along the coast from the Klasies cave towards Cape Town is a second occupation site, Nelson Bay cave, which was inhabited much later, less than 30,000 years ago. Here the toolkit is typical of the Later Stone Age, as the Upper Palaeolithic period is termed in Africa, featuring many refined implements cut from bone. The bones and other remains in the cave indicate that the Nelson Bay people caught and ate a greater variety of animals, birds and sea creatures than did the inhabitants of Klasies, including some dangerous animals that the earlier group had avoided. By studying the age range of the animal bones in the caves (see p. 60), Richard Klein is also able to conclude that they hunted in a much more systematic fashion than had been apparent previously. 'The people of the Middle Stone Age [that is, between 130,000 and around 35,000 years ago] were less effective at exploiting aquatic, terrestrial and bird food resources than the Later Stone Age people who succeeded them,' Richard Klein states.

The control of animals

Although the Later Stone Age people of Nelson Bay had undoubtedly extended the range of foods available to them, they left behind no clear indication of direct control of their food resources. What of the people of the Upper Palaeolithic in France? Are there signs of husbandry here? Some people claim that there are a few pieces of evidence which suggest systematic herding of prey animals. Firstly there are some living sites where only one prey species appears to have been hunted. In southern France, towards the end of the Ice Age, for example, the reindeer was the main prey animal. Reindeer are migratory animals and it is possible that early humans followed the herds much as the Lapps and modern Siberian people do. Whether these Ice Age people exerted any control over the herds is another matter, but it is interesting that many living sites are unsuitable locations for generalized, opportunistic hunting, but are superbly strategic for the manipulation and possible corralling of herds. Secondly, there are some engravings which can be interpreted as showing harnesses on horses' necks. Thirdly, there is a strange pattern of wear on horses' teeth that is today only seen in domestic horses. Let us examine the strength of each of these pieces of evidence.

Living sites from early on in human prehistory usually contain a wide selection of animal bones, implying that the people who lived there were opportunists, taking whatever they chanced upon. Later in the fossil record one finds striking accumulations of bones of a single animal species. Some of these, such as the site at Olorgesailie in Kenya where there are the remains of over sixty giant baboons, suggest nothing more than selective hunting. But the finding of accumulations of, for instance, ibex bones in Upper Palaeolithic France or gazelle remains in early Israeli sites, do suggest the possibility of some kind of husbandry.

Let us consider a site on the Vicdessos River, just 3 kilometres (2 miles) upstream from where it joins the Ariège, in the French Pyrenees. There, perched above a scree slope and below craggy cliffs is a cave known as La Vache, where ibex hunters lived 13,000 years ago. The cave is situated at a strategic vantage point at the bottleneck of this narrow, steep-sided valley.

The Upper Palaeolithic hunters would naturally have experienced much

The view over the Ariège valley from a spot close to La Vache cave. Situated at a vantage point above the narrow valley, caves like this were home to hunters who probably controlled the movements of ibex herds 13,000 years ago.

colder conditions than prevail there today, and there would have been fewer trees, bushes and flowering plants. 'This was ideal ibex country,' says Paul Bahn, who has made a study of many of the Pyrenean sites. It seems that the cave dwellers at La Vache exploited their chosen prey over many centuries, leaving behind them rich layers of food debris. Analysis of this food debris has shown that of the bones from large animals more than eighty-five per cent were from ibex. La Vache was a winter home for the hunters. 'The ibex came down from the higher peaks during the coldest time of the year to escape the snow,' Paul Bahn suggests. So too, it appears, did flocks of arctic grouse and rock ptarmigan, the bones of which also contribute to the cave's domestic litter.

Is there any reason to think that the inhabitants of La Vache were controlling the ibex herds? 'Ibex are extremely difficult to hunt,' Paul Bahn explains. 'A single hunter attempting to pick off animals in gregarious groups usually has little luck, and the herd is inevitably disturbed. Hunting in groups is more successful. The Palaeolithic people may have exploited the tendency of ibex to go uphill when driven, directing a small group from the main herd towards concealed hunters. I believe that the fact that the people at La Vache successfully hunted these animals over a long period of time must imply some degree of controlled manipulation.'

Not what one would call true pastoralism, but, if Paul Bahn is correct, then the people of La Vache and of several similar sites in southern France must have taken great care to separate off groups for slaughter from the main herd and to ensure that the herds never strayed too far from the cave sites. There is no suitable term to describe this form of animal management: domestication involves control over breeding, which the cave dwellers almost certainly did not have. But they may have been intruding far more into the animals' lives than is suggested by the term 'hunting'.

Towards the end of the last century a remarkable collection of horse bones, about 17,000 years old, was found near to the foot of a cliff at a small village called Solutré in east central France. Several thousand wild horses died at this spot. One explanation of this extraordinary accumulation of bones proposed that the hunters of the day stampeded horses so that they fell over the edge of the cliff to their death. However, the horses would first have had to be driven up a substantial slope to reach the cliff edge. And the animal bones are some distance from the place where it is suggested they met their end. Perhaps a more plausible explanation is that the hillside was used as a barrier against which to trap the animals in seasonal drives. The archaeologist who first studied the bones, Henri Toussaint, went so far as to suggest that the Solutreans had domesticated the horse, a suggestion for which he was roundly criticized when he published it in 1873. But, as with the bones at La Vache and other ibex sites, the Solutré horses at the very least give a definite impression of some form of human control over prey animals.

The 'Age of Reindeer'

A number of interesting sites in the Pyrenees and further north in the Périgord suggest that, as the Ice Age was nearing its end, reindeer began to assume prime importance. There are caves in which the food litter is made up largely of this single species, and this period has often been called the 'Age of Reindeer'.

Reindeer undergo long annual migrations, sometimes travelling several thousand kilometres. As British archaeologist Derek Sturdy has pointed out, if a population group opts for living wholly or partly on reindeer meat, they can either stay in one place and hunt the reindeer when their migrations bring them within striking distance, or they can follow the migrating herd. For a people who have alternative food resources for much of the year, there is the possibility of single-season exploitation. However, both Derek Sturdy and Paul Bahn believe that Palaeolithic hunters were far more likely to have adopted the strategy of herd-following.

Reindeer prefer a cold climate, and today are found only in the northernmost parts of Europe: in Norway, Sweden, Finland and Iceland. During the Ice Age, however, they were found throughout France and even in northern Spain. It seems likely that they migrated constantly, as modern reindeer herds do, always moving to cooler regions during the summer months.

If the reindeer hunters of the European Upper Palaeolithic did take to herd-following, what consequences would this have had? Obviously, they must have been extremely mobile and the human population as a whole must have been small and thinly dispersed. There are other implications too, as Paul Bahn explains: 'I'd be very surprised if the lifestyle of Upper Palaeolithic people of the Pyrenees and the Périgord was not very similar to that of some of the Siberian reindeer tribes. The way they conduct their reindeer economies varies considerably: some have complete control over

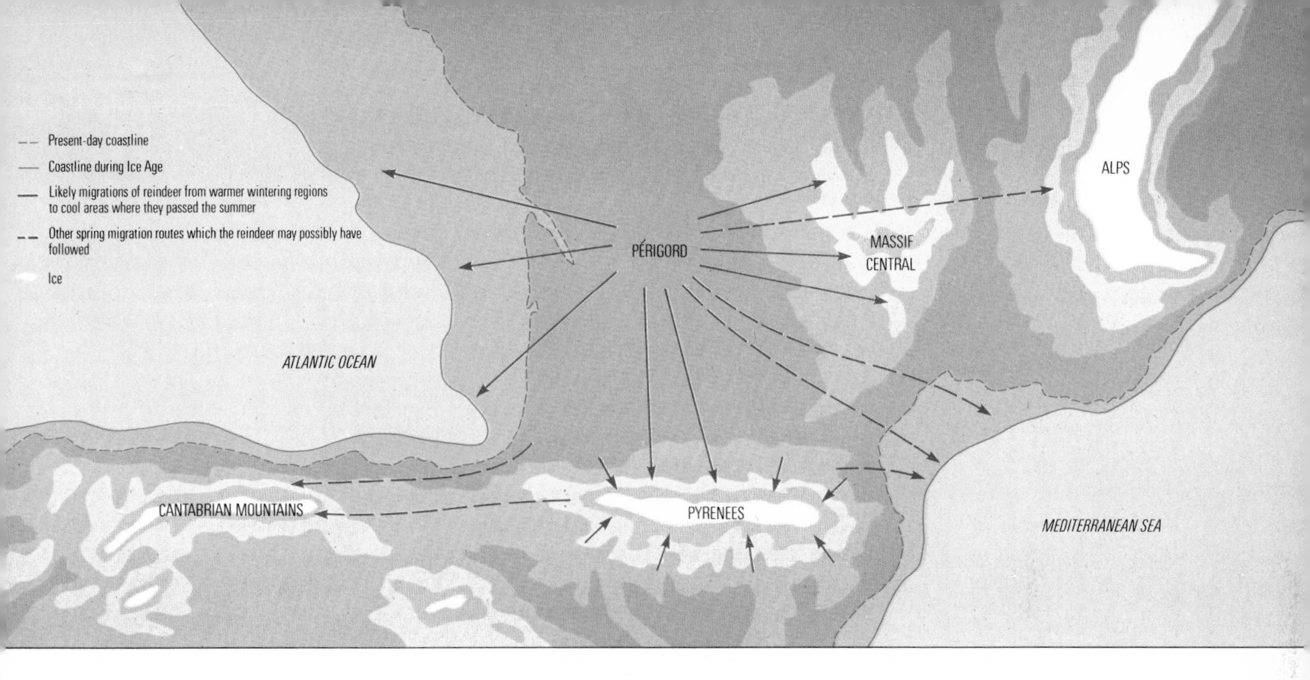

Many strands of evidence suggest that the reindeer and human populations of the Périgord moved to the Massif Central or the coastal plain during the summer, while those from the Pyrenean foothills moved to higher parts of the Pyrenees. There are other possible migration routes: to the Cantabrian mountains in the west, to the Mediterranean coast, or even to the Alps, but the evidence to support such migrations is more tenuous. The distance travelled to the coast would have been greater than it is today, the sea level being much lower because of the water locked up in the ice caps.

their animals; some hunt reindeer, but domesticate a few for milk and for traction; and some migrate with the herds while others are more sedentary; it all depends on local circumstances.'

It is certain that the caves of the Pyrenees and the Périgord were inhabited by reindeer hunters during some parts of the year in the late Upper Palaeolithic. But what can one say about the possible mobility of these people? Paul Bahn has looked into this question and has concluded that the encampments in these regions are seasonal. 'Throughout the Périgord sites there is a distinct lack of shed female antlers, a fact which suggests that the deer were absent during the summer.' The same holds for the caves in the Pyrenees: these too were occupied mainly during the winter.

Further evidence comes from a recent study by Arthur Spiess of reindeer remains from the cave of Abri Pataud in Les Eyzies. The bones, the antlers, the tooth-eruption and above all his new technique of sectioning the teeth and counting the layers of cementum all combine to show very clearly that the site was only occupied from late autumn to early spring. Where did the herds of reindeer that spent their winters in the Périgord and on the lower slopes of the Pyrenees move to during the warmer months of the year? Some people have suggested that the herds in the Périgord, for instance, might have travelled a mere 70 kilometres (45 miles) to the first slopes of the Massif Central. But the climate there would not have been all that different from that of their winter quarters. From the Périgord the reindeer herds would surely have had to go much further into the Massif Central or to the higher pastures of the Pyrenees to find cool summer pastures. Perhaps the Swiss Alps were the summer target of some of these herds, a journey that would have taken them more than 450 kilometres (280 miles).

According to Paul Bahn, however, there is good reason to believe that at least some of the Périgord reindeer herds headed for coastal grazing grounds in the summer, either to the Atlantic littoral or the Mediterranean, while others went to the Pyrenees. Meanwhile, some of the animals in the lower Pyrenean wintering valleys may have gone southwest in the summer, as far as the Cantabrian mountains. The direction of movement is of course determined to some extent by the physical barriers that may lie in the path of a migrating herd. One piece of evidence that Paul Bahn looks to in herd movement between the Périgord, the Pyrenees and the Cantabrian mountains is the striking uniformity of the art throughout these regions. This undoubtedly implies contact between people in these different areas, and it is possible that a small number of artists painted pictures in a large number of widespread sites.

The possible use of coastal summer grazing is indicated by finds of shells. Virtually all the reindeer caves of the Pyrenees and the Périgord contain sea shells, many of which are pierced as if to be worn. The type of shells found in caves depends on their locations. For instance, those in the Périgord and the western Pyrenees have a high proportion of shells that must have been brought from the Atlantic coast of Ice Age France, though there are also some Mediterranean species. However as one moves east, the proportion of shells from the Mediterranean increases. Some prehistorians have imputed high prestige value to these shell pendants and decorations, suggesting that the people of the Pyrenees and the Périgord made special journeys to the coast to collect them. But it is just as likely that the shells were collected while these Ice Age hunters were following the migrations of the reindeer.

There are occasional glimpses of the reindeer hunters' coastal visits in their art, with flatfish depicted on pieces of portable art found at the Lespugue caves in the Haute-Garonne and at the Abri du Souci shelter in the Dordogne. A marine fish resembling a sole is represented on a wall of the Mas-d'Azil cave. Taken with the other evidence, particularly that from marine shells, the notion of seasonal migrations of people in the wake of their reindeer herds seems very strong indeed. And, as Paul Bahn suggests, although it is not possible to know yet how much control the hunters had over their animals, they may have manipulated the herds in the same way as Lapps and modern Siberian people pursuing a similarly based economy.

An engraved fish from the Spanish cave of Altxerri, which is not far from the sea. Images of fish have also been found in some caves in central France, suggesting that their inhabitants spent part of the year at the coast approximately 300 kilometres (190 miles) away.

Domesticated horses in the Ice Age?

The last threads of evidence concerning the way of life of the people of Palaeolithic Europe relate to horses: firstly to putative bridles; and secondly to an interesting pattern of wear on some teeth. Both topics were discussed at the beginning of the century in French academic circles, but both suffered heavy criticism and temporary eclipse. It is only recently that discussion has been revived with some interesting new evidence.

At the end of the nineteenth century, Edouard Piette was the principal proponent of Palaeolithic domestication, both of reindeer and horses. By 1875 he was very sympathetic to the idea of ancient domestication after he had seen engravings on a piece of portable art from Laugerie-Basse. This appeared to show a male reindeer wearing a halter. A scattering of tantalizing images of this sort showing horse heads kept turning up and this kept Edouard Piette's interest in the idea very much alive. When a horse-head carving apparently showing a rope-like harness was found in 1893 in the cave of St-Michel d'Arudy in the western Pyrenees, he was fully convinced. However the powerful Abbé Breuil was unconvinced. At first Abbé Breuil had warmed to the evidence for such early domestication, but later he changed his mind and it was not until very late in his life that he was prepared to reconsider.

Edouard Piette died in 1906, after which time opinion was strongly influenced by Abbé Breuil's stance. There the matter rested until 1966, when two French prehistorians published a substantial paper that both re-examined the question and contained a new piece of evidence: an engraved horse head from a site in southwest France known as La Marche, which is between 15,000 and 14,000 years old. The engraved lines strongly suggest

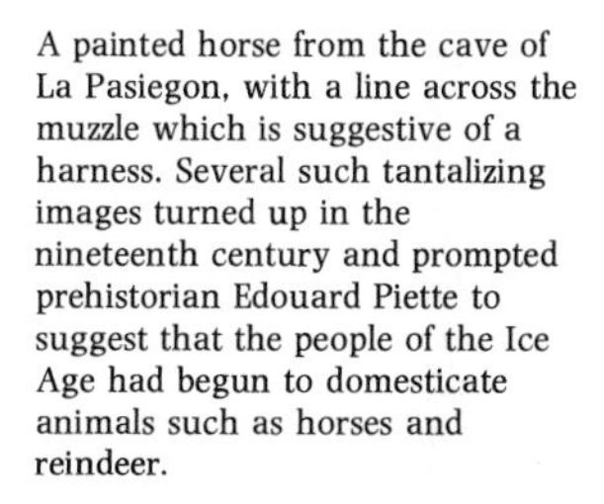

A painted horse from the cave of La Pasiegon, with a line across the muzzle which is suggestive of a harness. Several such tantalizing images turned up in the nineteenth century and prompted prehistorian Edouard Piette to suggest that the people of the Ice Age had begun to domesticate animals such as horses and reindeer.

that the animal was wearing a harness. It is unlikely that the lines represent stylized anatomy: the lines down the head, across the muzzle and under the chin were crafted after the rest of the carving was complete, and they form the perfect shape of a harness.

Paul Bahn, who has examined this topic in depth, puts his feelings this way: 'For the most part you can't be one-hundred-per-cent emphatic about the engraved harnesses: it's art, and art is always ambiguous. But the really clinching evidence is the engraving from La Marche. The lines cannot be confused with the horse's musculature. It has to be a harness; it was drawn on after the horse head was finished.'

The use of a bridle on horses such as the one seen on the La Marche engraving implies that people either rode the animal or used it for traction. 'I find neither of these possibilities particularly startling at this period, that is, around 14,000 or 15,000 years ago,' says Paul Bahn. 'Upper Palaeolithic people were of exactly the same intelligence as we are. You'd expect that it would dawn on them that they might be able to do more with horses than simply throw a spear at them when they were feeling hungry.' One is left to guess how Ice Age people made use of harnessed horses. The image of them galloping across the chilly grasslands of Europe may conflict with archaeologists' preconceptions about life at this period, but it may well be accurate.

Further support for the idea of bridles comes from one of the most enigmatic group of objects yet to emerge from the Palaeolithic, the so-called *bâtons de commandement*, or 'commanders' maces'. These strange implements are cut from bone or antler, they are pierced, usually once, sometimes twice, at one end, and are frequently engraved. As their name implies, they have been widely thought to be a sign of power and distinction. A more recent suggestion is that they were used for straightening the shafts of

The carving of a horse head from St-Michel d'Arudy. Edouard Piette was fully convinced by the rope-like harness which it appears to wear, but opposition from the influential Abbé Breuil prevented widespread acceptance of his theories.

arrows: the end of the arrow is placed through the hole, and the shaft is then bent against its natural curvature so as to produce a perfectly straight weapon. Some analysis of the holes has, however, shown that the wear around them was probably caused by a soft material rather than by wood or bone which would have been used for arrow shafts. Could these mysterious batons have formed part of a harness? Perhaps they had the rein passed through them and were held in the hand as a way of guiding the animal, in a similar way to objects used by Eskimo people today?

Last of all we turn to the horse teeth with which the chapter opened. Here the story begins once again in France, with an adventurous prehistorian who was prepared to say what he believed he could see in the evidence. In 1910 Henri Martin discovered some front teeth of a horse that showed every sign of a condition known as crib-biting. Stabled or tethered horses occasionally press their teeth onto some form of hard support – part of a door post or fence, for instance – arch their neck, and inhale deeply. No one knows what makes horses engage in this bizarre piece of behaviour, but most veterinarians agree that they do it through boredom. One consequence is that the horse's incisor teeth become worn down in an unusual way, so that instead of having squared-off edges at the bite-point they are chisel- or wedge-shaped.

A variation of this pattern of tooth wear is precisely what Henri Martin saw in the teeth he examined, but the horses to whom the chisel-shaped incisors had once belonged died at least 30,000 years ago: they were from a site in the French Charente, called La Quina. 'It was an incredible fluke that Henri Martin discovered the teeth in the first place,' Paul Bahn says. 'He was a good enough palaeontologist to recognize that it was crib-biting, he was broad-minded enough to accept that it might exist at this early period, and he had the courage to publish it.'

Luckily, one of Henri Martin's closest friends was an eminent French veterinarian who was keenly interested in archaeology. The veterinarian, E Hue, was later able to examine the teeth of 20,000 modern American horses. His conclusion, which he published in 1915, was that crib-biting is unknown in wild horses: it occurs only in animals that are regularly tethered for considerable periods of time. The implication was clear: humans at least 30,000 years ago *were* doing more than just throwing spears or firing arrows at horses.

Paul Bahn managed to find, in the Musée de St Germain, one of the worn teeth that Henri Martin originally discovered. But his trip to Paris was also rewarded with another, far more substantial piece of evidence. 'What I was looking for,' he now explains, 'was the whole front of a mandible, not just a single incisor. By a great stroke of luck I managed to find the front of a lower jaw intact showing clear signs of chisel-worn, polished teeth. I looked through hundreds of teeth in the museum's collection, and the incidence of unusual wear was very low. This makes me certain that this particular jaw displays something significant: if it's not crib-biting, I'd like to know what it is.' The site from which Paul Bahn's horse teeth were excavated originally was Le Placard, which, like La Quina, is in the Charente, but is somewhat younger.

What does this ancient crib-biting tell us about Ice Age practices? Paul Bahn is cautious: 'I wouldn't go as far as to say "horses tied up" equals "domestic horses". Everyone has his own interpretation of what domestication means. The animals must certainly have been tethered quite regularly to produce this form of behaviour, but perhaps they were used as tame decoys or simply as pets rather than being used for riding.'

The *degree* to which the people of the Upper Palaeolithic had control over

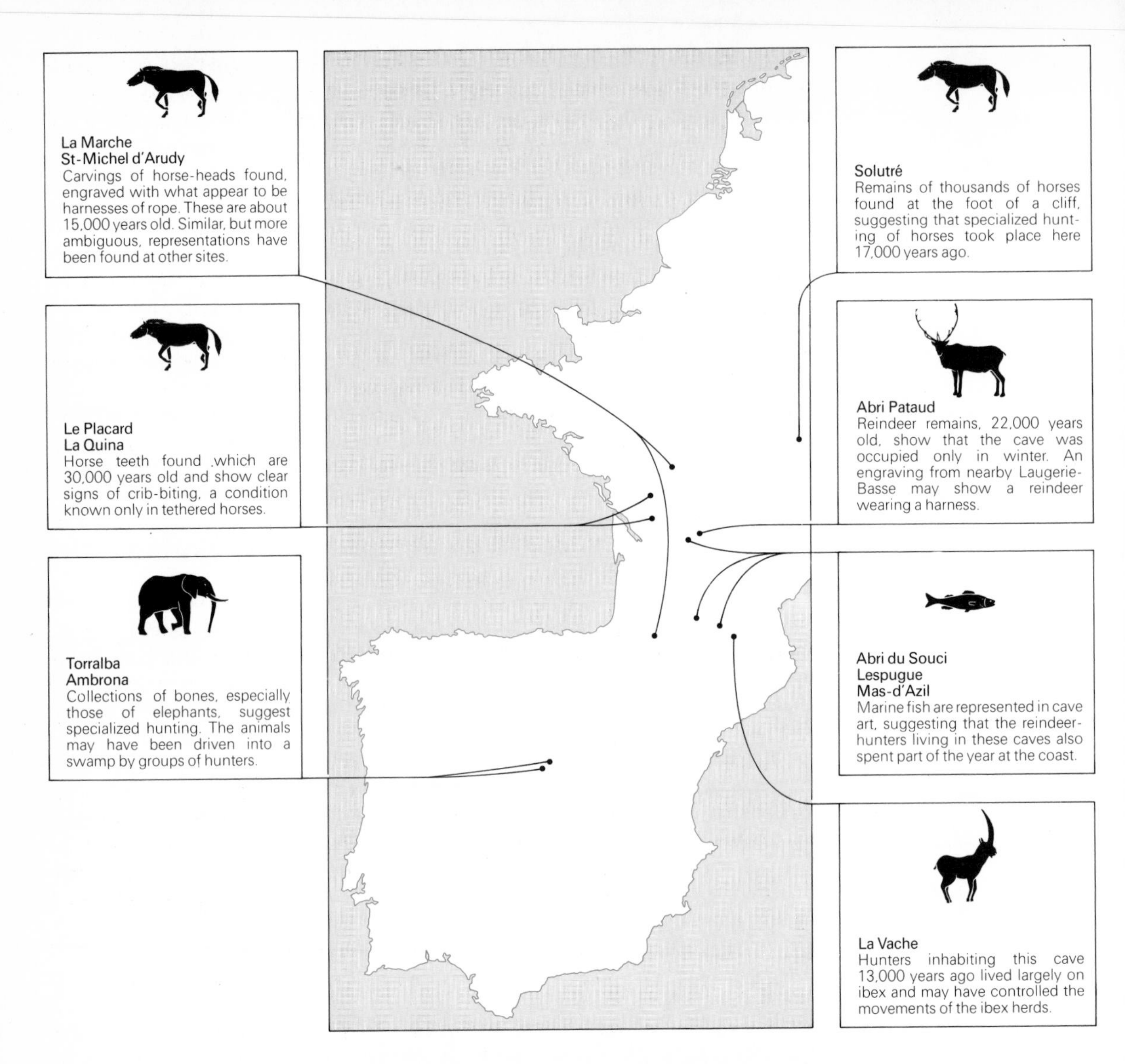

animals in their environment – horses, reindeer, ibex and so on – remains to be proved, but the evidence of husbandry now appears to be quite strong. Very probably, these people did not confine their skills to animals: 'I'd be very surprised if they weren't manipulating plants in the same way,' Paul Bahn says. Clearly, the archaeological record contains more clues about the style of life of Palaeolithic people than have previously been recognized. Whatever the final story turns out to be, it now seems inescapable that the transition from hunting-and-gathering to agriculture was not as sudden or as recent as archaeologists once believed.

New evidence from Africa

Recent evidence from Kenya further confirms this impression. It was once generally believed that cattle had been brought to East Africa 1,500 or 2,000 years ago by people who migrated from the north and who had acquired cattle, together with many other features of their culture, from civilizations then flourishing in Egypt and northern Ethiopia. But it now seems that there were domesticated cattle in Africa a great deal earlier than this. Charles

Many different strands of evidence support the idea that Ice Age people were learning to control the animals they hunted. Systematic hunting was probably the first step, and it appears that this began long before the Ice Age, among the *Homo erectus* populations of Torralba and Ambrona. The second step may have involved controlling the movements of the animal herds, as apparently happened at La Vache and Solutré. Finally, individual animals could have been tamed and used for transport as well as meat, as the finds of St-Michel d'Arudy, La Marche, La Quina and Le Placard suggest. The dates of the different finds imply that the degree of domestication varied a great deal from one locality to another.

Nelson from the University of Massachusetts has been seeking evidence for early pastoralism in Kenya's highlands since 1975. He reasoned that this environment would favour the introduction of domestic animals at a much earlier time than in the surrounding lowland regions, where there would have been tsetse fly and less favourable climatic conditions.

Over one hundred sites where prehistoric pastoral people had lived were found early on in the project. Test excavations at a number of these sites revealed that by 2,500 years ago there were not one, but *many* pastoral cultures. This degree of cultural diversity implied that pastoral peoples had been developing in East Africa for a fairly long period of time. However, the remains left by earlier pastoral people proved difficult to find initially.

One of the first finds was at a site known as Salasun, high on the volcanic crater of Mount Suswa where the Maasai still herd their goats and cattle. Excavations here showed that pottery, stone bowls and platters of the sort found in the more recent pastoral sites were everyday items some 7,000 years ago, and that domestic animals were used to supplement the hunting of wild game.

More recently Charles Nelson and his team have been excavating a number of sites near Nairobi at a place called Lukenya Hill. There is some evidence of cattle in the food remains at these sites and several specimens have now been dated by radiocarbon tests to an age of just over 8,000 years. Cattle remains at these sites constitute less than twenty per cent of the total animal remains which implies that they were not being kept in particularly large numbers at this early stage. About 3,300 years ago the picture changes and the finds show that people in this region began to be dependent on an entirely pastoral economy.

The question is *when* were cattle first domesticated and *where?* The absence of wild cattle south of the Sahara Desert throughout the fossil record implies that cattle were brought into East Africa as domesticated animals. It would therefore follow that cattle were domesticated somewhere outside Africa well before 8,000 years ago, and there is evidence for domesticated cattle in North Africa at a little over 9,000 years ago which lends support to this proposal.

In Europe, North Africa and Asia, cattle remains in early archaeological sites have generally been interpreted as those of wild cattle that had fallen prey to hunters. In tropical Africa, where there were no wild cattle, such an interpretation is impossible. It is very probable that some of the early archaeological finds of cattle in Europe have been misinterpreted, and that they are domesticated, not wild, cattle.

Faced with so many different strands of evidence, it is hard to escape the conclusion that the domestication of animals stretches farther back in time than we had previously imagined. It is clear that the story of pastoralism has not yet been fully told.

12 A New Way of Life

Coastal Peru, north of Lima, is a place of extraordinary contrasts. The coastal waters are enriched by the Humboldt current, which carries nutrients up from the sea bed to the surface, and the most prolific marine life in the world is found here. A few miles inland, mountain slopes reach up to the peaks of the magnificent Andean chain, and in the sheltered nooks of these slopes wild potatoes, beans and herds of llama thrive. Between the productive Pacific waters on the one side and the fertile Andean slopes on the other is the coastal plain, probably one of the most inhospitable stretches of land on earth. The arid white sand rarely feels the splash of rain, receiving only about 4.5 centimetres (1.75 inches) in twenty-five years. Nothing grows on the plain, apart from scatterings of a curious plant called tillandsia: looking like the severed top of a pineapple, it clings virtually rootless to the sand from which it absorbs the thin morning mist.

Yet this narrow strip of arid land has supported some of the greatest civilizations in the New World. The Inca Empire which eventually controlled an area greater than that governed by the Romans took root here. Before the flamboyant Incas were the equally artistic Chimu people whose state centred on the massive and orderly city of Chan Chan, sited just north of the river Moche. And preceding the Chimu were the Moche people, famous for their decorated pottery. The secret of success on the hostile coastal plain for these civilizations was simple: irrigation.

Every 20 kilometres (12 miles) or so the coastal plain is cut by rivers draining from the Andes and here swathes of green vegetation stand out brilliantly against the white sands. With irrigation modern farmers grow fruit and vegetables in some areas but 2,000 years ago farmers tilled almost twice as much land as is worked today. To the Chimu and Moche people, irrigation was a natural part of living, and they were immensely successful at it. A huge network of canals carried water throughout the Moche valley, and extraordinary engineering projects extended the irrigation system to neighbouring valleys. The Moche people were among the most accomplished early farmers in the New World, and were also among the most skilful of the world's early engineers.

It is not only ancient canals that sketch the history of coastal Peru in the desert plain. Monumental constructions rival the pyramids of Egypt in the organization and manpower that must have gone into their building, and it has become clear in recent years that many of these pre-date the well documented Moche state. In the Moche Valley alone there are at least six such sites which are more than 2,000 years old, and the nearby Supe Valley has ten sites with monumental architecture of this early era. Aerial surveys of the region reveal hundreds of ancient mounds in valleys up and down the coast, the vast majority of which are unexplored. Recent research by Michael Moseley and his colleagues at the Field Museum in Chicago has

The Chimu and Moche people built a vast network of canals across the Moche Valley, and added other canals to carry water into neighbouring valleys. This photograph shows an aqueduct on the canal between the Moche and the Chicama Valleys. Part of the aqueduct has been washed away and the structure can be seen in cross-section.

demonstrated that these structures are evidence of sophisticated populations living here as much as 4,000 or 5,000 years ago.

These ancient Peruvians were among the first in the New World to change from a nomadic hunter-gatherer life to a more sedentary existence, a change that was already under way in the Old World. In Peru, however, the transition was based primarily on the produce of the sea, not of the land. Communities in other parts of the world, such as the Pacific coast of North America and some coastal regions of Europe were also established on the basis of marine produce. Such developments simply indicate, of course, that where there are rich and exploitable resources, humans will exploit them.

The 'Agricultural Revolution'

The majority of the world's peoples adopted some form of agriculture and changed to a new way of life around 10,000 years ago. Why did the so-called Agricultural Revolution occur and how did it come about? The answer to the first question is the most difficult, and there is a striking lack of agreement between researchers. Many possible causes for the transition have been proposed over the years, some of which relate to external pressures on people, some to internal drives within people, but none of which, taken alone, is very convincing. The transition was probably the product of a combination of circumstances, some of which applied throughout the globe, while others were strictly local. The second question is easier because one can examine the archaeological evidence in order to reconstruct life in transition. The most impressive point to emerge is the diversity of approaches: people in different parts of the world went about the transition to agriculture in different ways according to local conditions, local resources, and, presumably, local traditions.

There are several fascinating aspects of the Agricultural Revolution. For one thing, the speed of the transition was remarkable. For perhaps two million years, human ancestors had practised nomadic hunting-and-gathering, a way of life that was characterized by stability rather than change in terms of technology and culture. Then the ancient way of life was virtually abandoned over a period of a few thousand years. It should be remembered that hints of the new way of life are to be found in the archaeological record as early as 30,000 years ago, which is 20,000 years before the date traditionally given for the beginning of the Agricultural Revolution. As I suggested in the previous chapter, the Agricultural Revolution is therefore best seen as a change in emphasis rather than a totally new invention. The speed of that change, nevertheless, was striking.

Equally striking was the accompanying change in world population. Biologists have estimated that the total human population of the world 10,000 years ago was between five and ten million. Within 8,000 years that figure had increased to three hundred million. During the next 1,750 years, that is to the beginning of the Industrial Revolution, there was a further increase of five hundred million. With the Industrial Revolution the modern population explosion began, bringing the current population level to around four thousand million, with a projected six-and-a-quarter thousand million by AD 2000. The later stages of this astounding growth were of course fuelled by the Industrial Revolution, but there can be no doubt that the Agricultural Revolution of 10,000 years ago gave the curve its initial upward trend.

Another interesting aspect of the transition to agriculture is that it occurred in several different locations quite independently of each other. There is some disagreement about the details, but Jack Harlan, of the University of Illinois, sees three major centres where agriculture developed

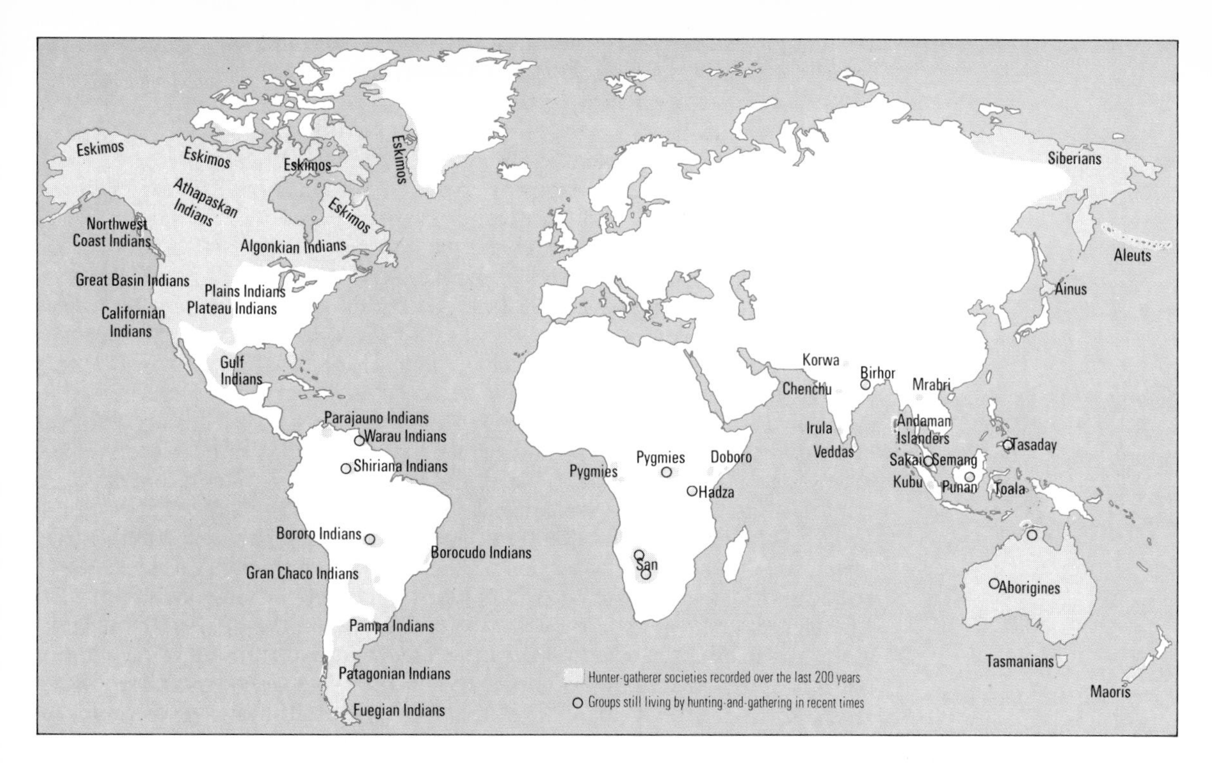

Most of the world's peoples adopted agriculture 10,000 years ago. With the exception of the Aborigines and the North American Indians, populations which remained faithful to the old hunter-gatherer way of life were small and scattered. They were often found in marginal, semi-desert environments, or in places where the forest was impossible to clear. Of those who were still hunting-and-gathering 200 years ago very few remain today, and many that have persisted with this way of life are being forced into settlement by government agencies. The nomadic hunter-gatherer lifestyle, which our ancestors practised for two million years or more, is now in danger of disappearing completely.

intensively: these are the arc-shaped area east of the Mediterranean (the so-called Fertile Crescent), northeast China and Mesoamerica. There are other, more diffuse areas of agricultural development in Africa, India, Central Asia, Southeast Asia and the northern and western regions of South America. As far as is known at the moment, the Fertile Crescent had the earliest agricultural development on a large scale, beginning at least 10,000 years ago. China followed around 7,000 years ago, and Mesoamerica came last, about 5,000 years ago. Although some people have tried to show that there was diffusion of agriculture between these major sites, the archaeological evidence supports the idea of independent development.

The speed of transition from a nomadic to a sedentary existence, the accompanying rise in population and the universality of the change are three major facets of the Agricultural Revolution, and they give some clues as to why the change occurred. Before looking at this question, however, it is important to set the event in its climatic context.

About 15,000 years ago the ice sheets began to shrink and warmer times were on their way. With the end of the Ice Age came a warmer and wetter climate, giving greater forest cover and richer grass and shrub growth. The sea level rose by as much as 130 metres (430 feet) as the ice melted: coastal plains disappeared under the ever-encroaching sea until one twentieth of the land surface had vanished. Ten thousand years ago the world was experiencing tremendous climatic change, and this may, perhaps, have been linked with the advent of agriculture.

A change in climate or a population explosion?

Anthropologist Richard Lee once described the hunting-and-gathering way of life as '. . . the most successful and persistent adaptation man has ever achieved'. It is also apparent that, in the early days at least, the shift from hunting-and-gathering to agriculture would almost certainly have involved a number of disadvantages: more time spent working, less meat in the diet and less security of food supply in the event of crop failure. Given all this, why did it happen at all? And, moreover, why was it such a universal phenomenon?

The universality of the phenomenon inevitably encouraged prehistorians to look for some single external factor that would have affected all the peoples of the world, and the most obvious candidate was a major change of climate. One of the early ideas, put forward in the 1950s by Gordon Childe, was that with the end of the Ice Age came a brief period of severe drought, at least in the Middle East. The humans, animals and plants became concentrated in the well watered valleys of, for instance, the Nile. Here, according to Gordon Childe, their close proximity to each other 'might promote the sort of symbiosis between man and beast implied in the word "domestication" '. He elaborated the idea, by suggesting that 'the hunters whose wives were cultivators had something to offer some of the beasts they hunted – stubble on grain plots and the husks of the grain. As suitable animals became increasingly hemmed in to the oases by the desert, men might study their habits and instead of killing them off-hand, might tame and make them dependent.' The problem with this 'oasis theory' is that no evidence has yet been found for the period of drought on which the theory rests.

Another theory proposes that the environment stimulated farming at least in the Near East, but points to the post-Ice Age spread of grasses which were ancestral to cereals as the key event. 'Ice Age hunter-gatherers were living in this area,' says Gordon Hillman of University College, Cardiff, 'and one of the things they were eating was seeds from stipa, the grass that grew well on the open steppe. As the world became warmer and moister, wild cereals spread from the sheltered woodlands and began to colonize the open steppe.' It should be pointed out that 'cereals' are simply those grasses which happen to have been chosen by man for cultivation. These grasses had larger seeds because they were adapted to growing in woodland, where moisture is plentiful but there is intense competition for light and space. A large seed produces a strong, fast growing seedling which can establish itself rapidly and compete successfully with the seedlings of other species. Gordon Hillman believes that the spread of these large-seeded grasses had an important effect on the hunter-gatherers of the Near East. 'When the cereals arrived, the people living there were already used to gathering seeds and would have readily exploited this rich new resource. Increasing exploitation eventually led to increased dependence.' According to Gordon Hillman, once the initial step had been taken, the path to active cultivation and domestication was set.

Richard MacNeish, of the Peabody Foundation for Archeology, Massachusetts, also sees environmental change as the key to the advent of agriculture in the Tehuacan Valley in Mexico. He sees the effect of environment in the following way: 'The warmer, moister climate in Mexico led to a spread of woodland and forest cover, and this reduced the number of potential prey animals available to the hunters there. They had to adapt their habits, and spend more time systematically searching for fruits, berries and nuts. Gradually they began to tend and nurture their wild food supplies, and eventually adopted substantial agricultural practices.' Richard MacNeish compares this process with the events which Gordon Hillman suggests took place in the Fertile Crescent: the food items are different, but the impulse for man's more extensive and systematic exploitation of them is similar, he argues.

Many other prehistorians, while agreeing that the new climate may

The wild ancestors of wheat were among the large-seeded woodland grasses which spread to the open plains as the climate grew warmer at the end of the last Ice Age.

have had a contributing effect, think that it cannot have been the only cause. Mark Cohen, of the State University of New York, recently wrote: 'In general, climate-based explanations suffer from two problems. First, climate phenomena are reversible and repetitive; they cannot account for a single occurrence of an event or process that has demonstrated very little tendency for reversal. Second, climate changes are by their nature regional in scope and often opposing in adjoining regions. Hence they are inherently incapable of explaining parallel economic trends over a broad geographic region.'

If not climate, then what else could have caused the almost universal and remarkably rapid change to agriculture? One of the most striking events following the Agricultural Revolution was the increase in population. Nomadic hunter-gatherers are extremely restricted in what they can carry around with them, and this applies to children as much as to possessions. Furthermore, women in these societies cannot go on food-gathering expeditions while carrying more than one infant, and so births tend to be spaced out to about one every four years, as we saw with the !Kung. Birth control by a number of methods, such as extended suckling, herbal medicines and infanticide, is common practice among nomadic people, and the population stays relatively constant as a result. When nomadic hunter-gatherers become sedentary, the restriction on births is lifted, and the population can rise. As agriculture can support more people per unit area than is possible with hunting-and-gathering, a population increase is virtually inevitable.

Some prehistorians, however, see population growth as a *cause* of the Agricultural Revolution, as well as an effect. The world of hunter-gatherers, they suggest, was straining to provide for a steadily rising population, until a new and revolutionary method of feeding the hungry mouths was discovered. Such an idea has been popular for some while, and has been put forward by several different people. Mark Cohen expounds one of the more recent and sophisticated versions in his book *The Food Crisis in Prehistory*. He proposes that the steady build-up of population pressure explains why the Agricultural Revolution occurred at about the same time in several different parts of the world.

Although some prehistorians are sympathetic to this idea, others disagree. British anthropologist Barbara Bender, for instance, insists that evidence to test Mark Cohen's model '. . . is in short supply, but where available it offers no support'. Barbara Bender suggests that some data in fact contradict the idea: 'In southwest North America, intensification and a commitment to agriculture coincide with a decline in population; and in the Central Valley of Mexico the ratio of population to productive resources remains constant right through a period of economic intensification.' Kent Flannery, an archaeologist at the University of Michigan who has worked in the Near East and in Mexico, also remains unconvinced by the population model: 'I don't see any evidence anywhere in the world that suggests that population pressure was responsible for the beginning of agriculture. The population level is not remotely high enough to force anyone to do anything. In Mexico, for example, the population when agriculture began was unbelievably low, too low to put any kind of strain on the environment or on people.'

When researchers disagree so strongly about the interpretation of available evidence it is usually a clear indication that there simply is not enough evidence available. Until much more information is uncovered, the proposition will simply have to remain an idea.

Cultural explanations for the change to agriculture

What of explanations that look within the human being, rather than at external circumstances? In the 1960s, American archaeologist Robert Braidwood invoked gradual cultural progression as the initiator of the Agricultural Revolution. 'In my opinion,' he said, 'there is no need to complicate the story with extraneous "causes". The food-producing revolution seems to have occurred as the culmination of the ever-increasing cultural differentiation and specialization of human communities. Around 8000 BC the inhabitants of the hills around the Fertile Crescent had come to know their habitat so well that they were beginning to domesticate the plants and animals they had been collecting and hunting. At slightly later times human cultures reached the corresponding level in Central America and perhaps the Andes, in southeastern Asia and in China.'

Robert Braidwood saw our evolution as progressing through a number of identifiable levels. At first there was free-wandering hunting with increasingly complex tools. This was followed by an era of food-gathering in association with hunting. Next came an era of selective hunting, seasonal collecting, and a refinement of stone-tool traditions. This developed into a season-bound economy combining food-collecting and intensified hunting. Under certain special circumstances the food-collecting became highly systematic, permitting a semi-sedentary settlement. Fully-fledged village-based farming emerged after a period of opportunistic domestication of available plants and animals. Robert Braidwood sees agriculture as offering significant economic advantages, so that, once the appropriate level of technological skill had emerged, the acceptance of the new form of economy would have been automatic. The transition could not have occurred earlier, he argues, because the cultural evolution of *Homo sapiens* had not reached the required stage.

Human cultural and economic attainment has progressed steadily through history, this is true, but how is one to distinguish cause and effect? Might not the new technologies associated with agriculture be the result of applying established skills to a new form of economic activity, rather than being the cause of that activity? It is very difficult to see how one could prove this one way or the other.

One of the newest theories about the Agricultural Revolution suggests that 'technology and demography have been given too much importance in the explanation of agricultural origins; social structure too little'. This is how Barbara Bender introduced a recent scientific paper in which she argued that the seeds of the Agricultural Revolution are to be found in the elaboration of complex social systems among late hunter-gatherers. She suggests that societies became much more elaborate with the evolution of fully modern man *Homo sapiens sapiens* 40,000 years ago, and points to contemporary hunter-gatherers to show the importance of establishing alliances between tribes, between bands within those tribes, and within the bands themselves. Such alliances often involve the exchange of goods and sometimes of food. Hierarchies within groups may arise, and again these may be demonstrated by the distribution of food. Here, she suggests, one sees the beginnings of a need to produce surplus goods.

Once the amount of food needed for distribution and exchange reaches significant proportions, mobility at certain times of the year may become difficult. But when hunter-gatherers stay in one place they soon exhaust the food resources of the surrounding area. There would then be pressure to intensify food production, by whatever means, so as to sustain its members during this time. Intensification of food production could eventually have led to domestication and agriculture.

What evidence is there for this theory? There are hints of some degree of non-egalitarianism within hunter-gatherer societies going back almost 100,000 years: there are burials that have the appearance of status markings, for instance. Trade of materials such as obsidian throughout the Near East as early as 35,000 years ago, and perhaps of sea shells within southern France at a slightly later period, hint at the type of alliance systems invoked by Barbara Bender. Archaeological information from pre-agricultural North America indicates relatively settled winter camps, status burials, and trade connections over many hundreds of kilometres. The notion is interesting for at least two reasons. Firstly, it relates the advent of agriculture to our cultural evolution, and secondly, it blurs the traditionally sharp dividing line between the pre-agricultural and agricultural stages in our history. The social/cultural theory offers a much more natural transition in mode of food production than is normally implied by the term 'Agricultural Revolution'.

Agreement on the question of why the Agricultural Revolution occurred is far away. Much more evidence is needed before outlining firm theories, and it may be that the Agricultural Revolution was so complex, and involved so many different factors, that no single simple model can ever describe it.

Let us turn to the question of *how* the change occurred. How did previously nomadic groups of people become sedentary people, pursuing an intensive food-production economy? The answer is that this varied depending on where people lived and therefore what resources were naturally available to them. In the Near East for instance, primitive forms of wheat and barley formed the new staple foods, backed up by the domestication of sheep and goats. Maize (corn) formed the backbone of the emerging agricultural economies in the New World, combined with beans which balanced the nutritional characteristics of the maize. The early Chinese farmers concentrated on millet as their staple with rice and soya beans added later on.

The early agriculturalists did not rely on just one or two crops. They continued to collect wild plant foods of different kinds and later domesticated a wider range of crops. In the Near East seeds of wild legumes (the bean, pea and lentil family) and seeds of wild grasses were gathered, as were pistachio nuts, almonds and capers. Later, crops such as oats, lentils, chickpeas, horse beans and linseed were domesticated.

The Fertile Crescent

I would now like to look in more detail at the events in the so-called Fertile Crescent, that arc of fertile land sheltered by the mountains and foothills of Israel, Jordan and Syria to the west, Turkey to the north, and Iran to the east. The area is covered with ancient settlements: Uruk, Ur, Jarmo, Ali Kosh, Abu Hureya, Jayonim, Jericho and many, many more. Here, 10,000 years ago, people were already engaged in intensive food production: village agriculture was an established way of life.

The Near East escaped the worst rigours of the Ice Age, but it was cooler, drier and more sparsely vegetated than it is today. Stands of oak and pistachio huddled in sheltered spots close to the Mediterranean coast; the wild ancestors of wheat and barley grew in limited stands among the trees. Hunter-gatherers exploited what they could in these sheltered spots, and hunted gazelle and oryx out on the steppe. They gathered the seeds from the stipa grass, perhaps by beating them into a container slung around the neck, or by simply uprooting the plants.

With the onset of a warmer climate the environment changed, not instantaneously or dramatically, but steadily and inexorably. Tree and bush cover began to spread, and so too did the wild ancestors of the cereals.

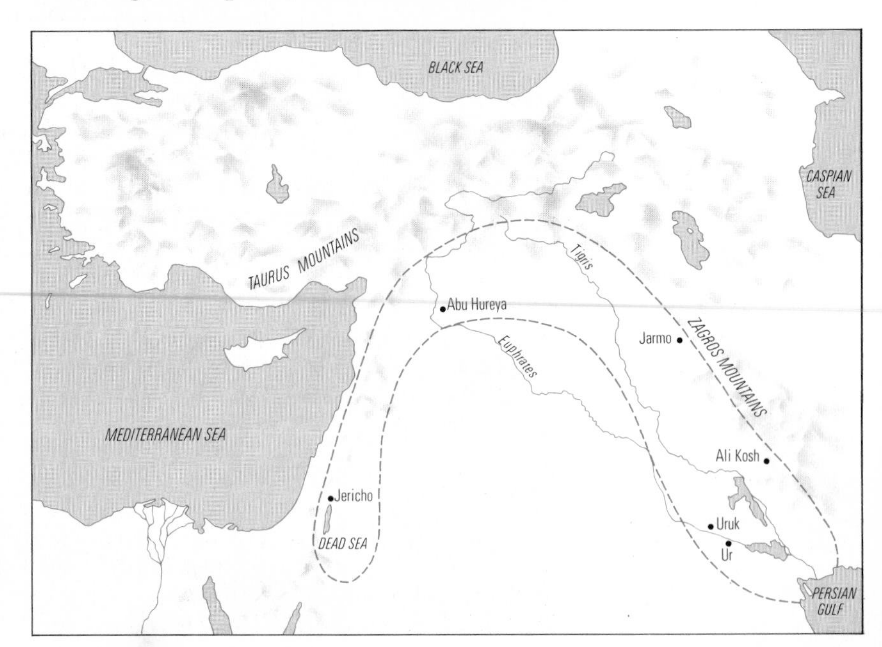

In the so-called Fertile Crescent, denoted by the shaded area on the map, village agriculture was an established way of life 10,000 years ago. This arc of rich, sheltered land is covered with the remains of ancient settlements.

Overleaf: An artist's reconstruction of the scene, some 10,000 years ago, as the wild cereals are harvested near a village in the Fertile Crescent.

Generation by generation the relatively sparse steppe was encroached upon by the advancing wild wheat and wild barley. As the millennia passed, thick stands of the cereal grasses filled out a previously sparse landscape. The hunter-gatherers must have welcomed the opportunity of including the larger cereal seeds in their harvest from the plains, but did the cereal-covered plains offer new possibilities, a new way of life?

In order to test this idea, Jack Harlan visited the slopes of a volcanic mountain in eastern Turkey called Karacadag. There, in the summer of 1966, armed only with a 9,000-year-old flint sickle, he began scything through the dense stands of einkorn, a variety of wild wheat. Within an hour he had harvested almost 3 kilograms (6 pounds) of grain. Even without a sickle he managed to gather 2 kilograms (4.5 pounds) though the process was very rough on his hands. Jack Harlan concluded that 'a family group . . . working slowly upslope as the season progressed, could easily harvest wild cereals over a three-week span or more and, without even working hard, could gather more grain than the family could consume in a year.' This simple experiment demonstrated that the post-glacial spread of wild grain had brought with it the potential for a sedentary life. People could gather the grain and store it for use when other foods in the area were limited. This would be permanent village life but *without* agriculture as such.

There is evidence that this actually occurred from the site of Abu Hureya in the Euphrates Valley of northern Syria. Andrew Moore, of the Pitt Rivers Museum in Oxford, led an emergency excavation to Abu Hureya in the early 1970s, when a dam project on the Euphrates threatened the site with

Using an ancient sickle such as this, Jack Harlan convincingly demonstrated that the fields of wild wheat and barley which grew in the Fertile Crescent were bountiful enough to allow settlement in one place. The people who began the move to agriculture here were probably still hunter-gatherers but they were no longer nomadic and they must have stored wild cereal grain to see them through the winter months.

flooding. The site turned out to have been occupied over two periods, once from 9,500 years ago to around 7,500 years ago, and once at the earlier, and more intriguing, time of 11,500 to 10,500 years ago.

Abu Hureya was a small village made up of simple pit dwellings, each of which had a framework of upright posts. Grinding stones, small flint blades and sickles, and a large number of bone implements, including needles, were recovered from the site. Judging by the accumulation of bone refuse, the people of Abu Hureya caught and ate rabbits, gazelles, sheep, goats and onagers (the wild asses of the region). The presence of freshwater mussel shells and fish vertebrae indicate that the people also fished in the nearby Euphrates. Most interesting of all, though, were the signs of plant foods: wild lentils, terpentine nuts (similar to pistachio), hackberries, caper berries, the seeds of feather grasses and, significantly, cereal grains.

Gordon Hillman examined these cereal grains and identified three types: the most numerous was the primitive wheat, einkorn, and the other two were wild barley and rye. The einkorn at the village, according to Gordon Hillman, shows no physical signs of domestication: the size, shape and other characteristics of the seeds are indistinguishable from those of wild einkorn.

One of the most significant differences between wild wheat and domesticated wheat is in the attachment of the seed to the rachis, or upper part of the stem. In wild wheat the rachis is very brittle once the seeds are ripe and, especially if the weather is warm and dry, it shatters readily, scattering the seeds. In a wild plant this is an advantage because the seeds fall to the ground where, when the autumn rains come, they can germinate and grow. But what is in the interests of natural seed dispersal is against the interests of the farmer. If he is to collect the grain by cutting the stems (rather than beating the grain off the stems into a container) he must do so at exactly the right moment, when the grain is ripe enough to be of use, but before the rachis has become so brittle that it shatters as he tries to reap the plants. Domesticated wheats differ from wild wheats in having a tough rachis which keeps the seeds attached to the plant even when they are fully ripe.

In wild wheat a very small proportion of plants have a tough rachis as a result of a genetic mutation. If early farmers routinely reaped their wheat they would inevitably lose some of the normal seeds, but all of the seeds from plants with a tough rachis would be collected. Thus, the proportion of seeds with the genetic mutation would be higher in the harvested seeds, and the mutation would be passed on when the seeds were sown in the next season. After many generations – Gordon Hillman has calculated this to take approximately 1,000 years – the proportion of the tough-rachis mutant would have risen sufficiently for an observant farmer to notice it. He could then take positive steps to select seeds of this more favourable variety for sowing, and the process of active domestication would have begun.

The fact that the cereals of Abu Hureya have a brittle rachis and generally show no physical signs of domestication suggest that the people who lived there were not farmers at all, merely sedentary hunter-gatherers. These people must have gathered wheat as it grew near to the flood plain of the Euphrates, and gathered it in such quantities that they were able to establish a permanent village base from which to exploit the other abundant food items of the region. Or did they? Andrew Moore is not satisfied that the people of Abu Hureya have been shown to be mere gatherers of the wheat rather than cultivators of it. Cultivation, he says, disturbs the ground, and this leads to the appearance of characteristic weeds. Evidence of such weeds, he points out, is found among the debris at the ancient village site.

It is, of course, possible that the inhabitants of Abu Hureya were in the early stages of farming, and that they had been breaking the ground and sowing seeds long enough to encourage the growth of certain weeds but not long enough to have had any effect on the physical characteristics of their crops. The evidence in favour of extensive cultivation is slender, and the simplest interpretation of this site is that if the people were indeed cultivators, then it probably represented only a limited part of their economy.

The transition from systematic gathering of wild wheat by a sedentary community to the cultivation of wheat would be easy and natural, should the circumstances demand it. This transition may have happened in many communities within the Fertile Crescent some 10,000 years ago, each community developing first the capacity for food storage that would enable it to settle in one place and then the skills of farming. As the community grew, craft specialization would emerge strongly, potters, textile-makers, tool-makers and basket-makers would have devoted their skills to the needs of the community, perhaps in exchange for food and other items. The small village settlement was then on the way to becoming a town.

A small village near Hebron, hardly changed in appearance from the ancient settlements of this region. Once the hunter-gatherers had settled in permanent villages and learned to store the wild cereal grains, the transition to true agriculture could have occurred quite easily.

Trade has a long history, but with the growth of sedentary communities, it would surely have expanded, both in volume and in the nature of items exchanged. As a natural consequence, trading centres arose, such as Jericho, north of the Dead Sea, and Catal Huyuk in eastern Turkey. Catal Huyuk goes back more than 8,000 years, and Jericho is believed to be the world's first city, established about 10,500 years ago. Both Jericho and Catal Huyuk grew into substantial centres, Jericho being surrounded by a high wall from very early times, and within each town social hierarchies became ever more clearly established and defined. While such transformations were taking place the newly developed techniques of food production were beginning to spread from the Fertile Crescent into Europe.

How did the migration of agriculture come about? Was it a movement of ideas, perhaps along active trade routes? Or did the farmers themselves move out and colonize new areas? Archaeological remains offer little help in answering this question: artifacts can be transported just as easily through trading contacts as by the migration of groups of people. But the genetic make-up of contemporary populations can reveal something about ancient migrations, as Luigi Cavalli-Sforza and his colleagues at Stanford University in California have shown.

By examining the distribution of certain genetic traits between geographically separated communities, it is possible to estimate how much contact there has been in the past between those people. For example, if population A has genetic marker *a* and a nearby population B has a genetic trait *b*, but there is no trace of *a* in population B and no trace of *b* among population A, then it is fairly certain that there has been little contact between the two populations. If, however, population A had marker *a* and a little of marker *b*, but population B had only *b*, then one could say that there had been migrations from B to A, but not from A to B. This, basically is how Luigi Cavalli-Sforza approached the problem of the transmission of agriculture from the Near East to the rest of Europe.

By measuring the distribution of thirty-eight different genetic markers in modern populations throughout Europe and the Near East, the Californian researchers produced a picture of the past that, though not entirely sharp, is unequivocal: it was the people who moved, taking their new ideas with them. However, one should not envisage a sudden invasion: the spread of agriculture took place at a rate of about one kilometre (half-a-mile) a year.

Agriculture in the New World

What of the New World? Agriculture arose here independently of the Old World, and the principal area was Mesoamerica, specifically the hills and valleys of Mexico. Domestication also occurred in various parts of South America, particularly in Peru, Ecuador and Bolivia. Kent Flannery describes some of the developments which took place: 'Domestication of plants in the New World is a very old phenomenon. For instance, people have been growing bottle gourds for at least 10,000 years, possibly longer. But it wasn't until around 5,000 years ago that agriculture in the full sense emerged, and it wasn't just agriculture as such, it was a real commitment to maize agriculture. This involved land clearance, intense labour investment, developing storage facilities and creating irrigation systems.'

New World hunter-gatherers, the indigenous Amerindians, were fortunate in having bottle gourds naturally available to them. When properly dried, the gourds make excellent containers, and the seeds, when toasted, are tasty and nutritious. 'These people must have planted bottle gourd seeds in strategic parts of their hunting-and-gathering territory,' Kent Flannery suggests. 'The gourds would then be available wherever they were needed.

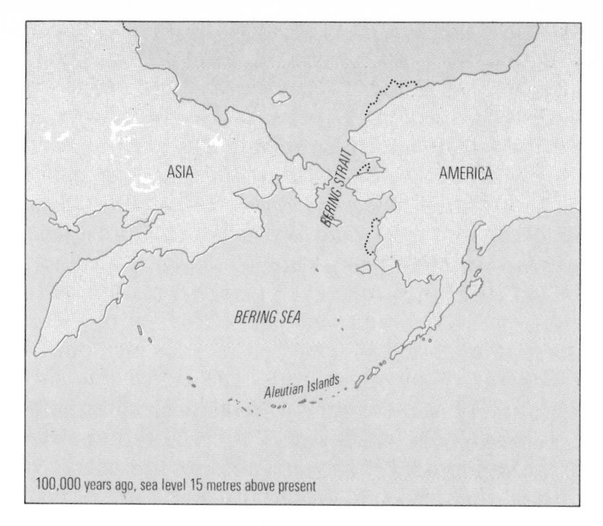

100,000 years ago, sea level 15 metres above present

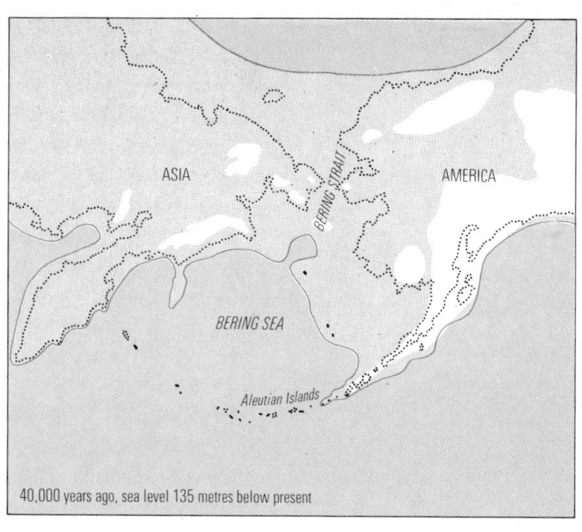

40,000 years ago, sea level 135 metres below present

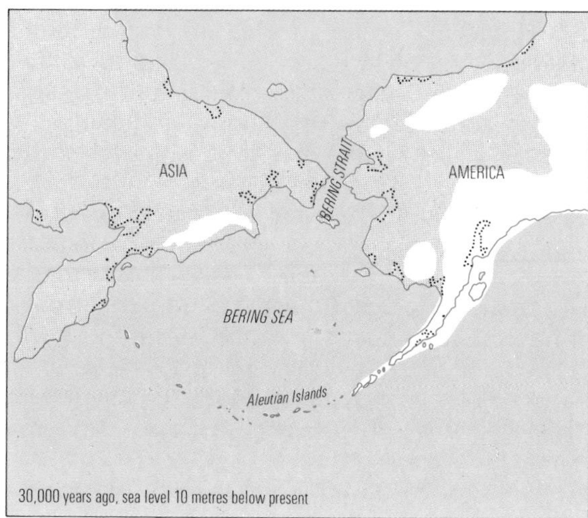

30,000 years ago, sea level 10 metres below present

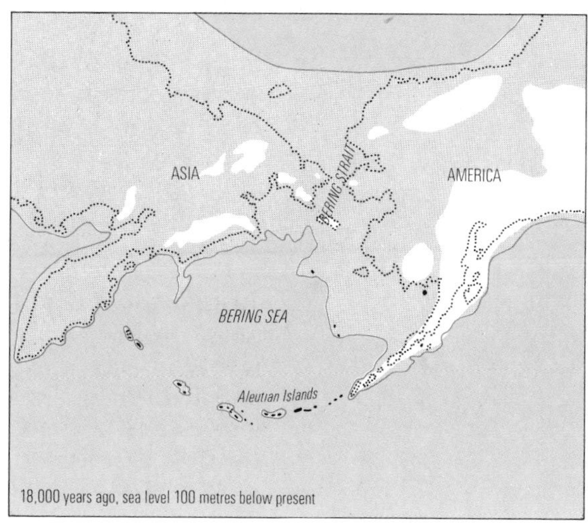

18,000 years ago, sea level 100 metres below present

Ice cap
Present-day coastline
Coastline then

Human populations penetrated the Americas during times of intense glaciation, when the sea level had fallen sufficiently to drain the Bering Straits. They may have entered the new continent while in pursuit of herds of game on the Siberian tundra. The first wave of immigrants probably arrived 40,000 years ago, but it is thought that there were further influxes later on, when the straits were once again dry land. One such influx may have occurred 18,000 years ago. The immigrant peoples gradually moved southwards, reaching Central and South America by at least 20,000 years ago, and settling into an agricultural way of life there by about 5,000 years ago.

And people would have recognized the similarity between the seeds of bottle gourds and related plants, such as squash [marrow-like vegetables], pumpkins, and many others. So, what started as casual cultivation of gourds for technological reasons, could have led to an extensive practice of planting for supplementary food items in the form of seeds. Of itself, this form of limited cultivation obviously isn't sufficient to support a sedentary community.'

The Amerindians were developing the cultivating habit at a time when the world climate was steadily warming following the Ice Age. During the Ice Age, Mexico was cooler and drier than it is now. The plains and valley floors were open grassland where antelope and jackrabbit lived and were hunted. With the shift to a milder climate thorn bush and forest encroached on the previously open spaces, providing habitats for the white-tailed and cotton-tailed deer at the expense of the previous inhabitants. The character of the environment undoubtedly changed. 'People must have had to make some alterations in their subsistence patterns,' suggests Kent Flannery, 'but whether this is enough to explain the eventual adoption of agriculture is open to question.'

It is, however, just such a change of climate that Richard MacNeish sees as being instrumental in changing the lifestyle of the Tehuacan inhabitants to sedentary agriculture. Tehuacan, an extensive valley in central Mexico is today intensively irrigated and farmed. Richard MacNeish and his colleagues have, over a period of many years, carried out thorough excavations and surveys in the valley, and their finds illustrate changes in the area over the past 12,000 years.

At the beginning of the period small numbers of hunter-gatherers lived in the valley, deriving almost seventy per cent of their food from game. Their life was ruled by the seasons. Meat was available all year round and was probably the only food to be had in winter. Pods and new plant growth became available in spring, seeds added nourishment to the summer diet, while fruit was eaten in the fall. In all seasons, the Tehuacanos ate opuntia and agave leaves to supplement these other foods. Life centred mainly around small family groups, 'microbands' as Richard MacNeish calls them.

The spreading bush and forest cover, nurtured by the warmer, moister climate, steadily reduced the amount of meat available to the people of Tehuacan. By 8,000 years ago, the meat had dropped to only fifty per cent of their food intake. In addition to the seasonal round of collecting pods, seeds, fruits and so on, the Tehuacanos began growing small quantities of squash, amaranth and chilli, making up perhaps five per cent of the diet. 'Valley people were moving about in well-regulated cycles so that the bands became tied to territories and, gradually, as they found abundant foodstuffs at particular places in the wet seasons, small groups coalesced into larger groups for brief periods.'

As the warming trends continued, so too did the evolution of the Tehuacanos' subsistence and social patterns: meat consumption continued to fall, seasonal food quests became more regimented, agriculture contributed more to the food store, and people spent more time in large groups. Cultivation of fruits might have arisen initially from the nurturing of wild stands of fruit-bearing bushes and trees, by clearing other plants from around them and preventing creepers from growing on them. Fully fledged horticulture arrived when people took to planting the seeds, and this occurred at least 6,000 years ago.

Agriculture was firmly established by 5,000 years ago, and substantial amounts of corn, beans, squash and chilli were being grown. Semi-permanent hamlets began to be established, and the trend to a completely sedentary life supported principally by agriculture had begun. By 3,000 years ago Tehuacanos grew almost half their own food, they lived in permanent villages and they were experiencing the problem of a population explosion. Signs of social hierarchies, cults, and regional political systems followed. Was the process initiated by a change of climate and therefore of environment as Richard MacNeish suggests? The idea is very attractive here, but there were surely other factors at work too.

The Tehuacan story must have occurred many times in many different locations throughout Mesoamerica. But there were variations, and one interesting one is to be found in the Oaxaca Valley 200 kilometres (125 miles) south of Tehuacan in the southern highlands. Kent Flannery has been excavating there, and he chose the area specifically to compare the developments with those at Richard MacNeish's site. 'The Oaxaca Valley is much more moist than Tehuacan,' explains Kent Flannery, 'which means that agriculture would have been easier to develop in the past, but the archaeological preservation can't be as good.'

Progress in the two valleys appears to have run pretty much parallel between 10,000 and 3,000 years ago, but then events took startlingly different courses. 'The minute there was a commitment to maize agriculture,' says Kent Flannery, 'Oaxaca just took off. It outstripped Tehuacan in terms of population, social complexity and social evolution. Within 1,000 years the valley went from having a scattering of villages to being dominated by a substantial city, Monte Albán, which covered 70 hectares [175 acres]. The city is believed to have been the first major urban centre of its type in the New World and it was innovative in both cultural and technological ways. The rulers of Monte Albán came to exert political power over a very wide area, often using violent tactics to achieve their ends.' Perhaps the environmental conditions of Oaxaca were a great deal more favourable than comparable sites, giving rise to a spectacular social development based on maize agriculture. 'It's the speed of the progression that amazes me,' says Kent Flannery, 'and it's bound to make us think about what's involved in the evolution of such systems.'

Let us return now to the part of the world with which this chapter began, coastal Peru. The coastal plain is rich in history, both in recent times and in the period when agriculture was undergoing its slow birth in the New World. But one of the interesting aspects of early Peruvian history is the development of highly sophisticated communities that were *not* supported by agriculture.

About an hour's drive south of the coastal town of Trujillo is a location known as Salinas de Chao, which lies 3 kilometres (2 miles) from the sea. A mound 25 metres (80 feet) high marks the centre of what, 4,000 years or so ago, was a thriving community. Terrace platforms and foundations of other structures are half hidden by sand and grass. 'There's no ready source of vegetable food in the area, such as roots, fruits, berries and so on: it's too dry,' explains Michael Moseley. 'And there are no suitable animals for domestication. It was much the same when people lived at this site. So how did they survive? The answer is to be found by running your hands through the sand here: anchovy bones by the thousand; bones from deep water fish; remains of shellfish. The people of Salinas de Chao made a living from the sea.'

One should not think of a small party of people spending their time fishing and existing as a simple coastal community. Moseley and his colleagues are now certain that Salinas de Chao supported a sizeable population, perhaps as many as 2,000 people, and that the site was occupied for

A stone frieze from Chan Chan, centre of the Chimu civilization which preceded that of the Incas in Peru. The frieze shows men fishing, a significant activity for these people, whose forefathers abandoned hunting-and-gathering to settle by the coast and live off the fruits of the sea. In time they turned to agriculture, moving inland and irrigating vast areas of the coastal desert.

Peruvian fisherman of today, setting out in simple boats made from reeds which they cultivate in sunken gardens near the coast. The people of Salinas de Chao probably used similar craft to fish these rich waters 5,000 years ago.

many hundreds of years. Very little remains of the domestic buildings, but the mound speaks for the state of the social organization. 'Although we should try to avoid reading too much into such structures – imposing preconceptions from what we see in other parts of the world for instance – it's clear that this mound represents at the very least a substantial amount of organized effort. The people who lived here were part of a complex social structure, almost certainly with a well defined hierarchy. Concerted effort towards bringing in food for a community is one thing. Communal work on a mound of this sort – probably some show of authority – is quite another.'

Salinas de Chao is not an isolated example: there are dozens of similar sites up and down the coast. They all indicate that where there is a sufficiently rich food resource, large sophisticated communities can evolve, and that agriculture as such is not the only possible basis for large settled communities. Not that these people ignored the land. Indeed, they successfully cultivated a number of crops, particularly bottle gourds and cotton, both of which provided essential raw material for fish nets: the gourds for floats and cotton for the mesh. It would be surprising if these coastal people did not eat the gourd seeds, and it is clear that they grew squash too. But the fact remains that their principal livelihood was in the sea.

Coastal Peru is very active geologically: the Andes are still being thrust upwards and at the same time the beaches become tilted away from the sea. As a result, the shoreline retreats at intervals, leaving beaches stranded in dry steps. The arid sands of the coastal plain shimmer as the sun strikes the shell fragments of ancient shorelines. For a people dependent upon marine resources, a retreating sea is highly undesirable. All the pre-agricultural coastal sites in Peru are now to be found several miles inland, separated from the sea by dry stretches of sand that once were the basins of shallow bays. For this reason, possibly combined with a general deterioration of climate, or perhaps even following a political edict, people moved inland, abandoning their fishing but taking their knowledge of cultivation with them. Agricultural settlements sprang up in the coastal plains and valleys. Impressive engineering skills created the beginnings of massive canal systems. Monumental architecture, which had its beginnings at the coast, matured into gigantic and spectacular forms. The signs of post-agricultural human potential were appearing: it was a potential that was being realized throughout the globe.

13 The Making of Human Aggression

A lucky discovery in 1937 brought to light one of the most arresting products of early Andean civilization. Julio Tello, Peru's pioneering archaeologist, was prospecting for historic sites along his country's coast north of Lima in June of that year, and he chose to rest overnight at a small hacienda in the village of Sechin. While he was there he saw a curiously carved stone that had been brought to the house as an ornament. Thinking that the carving might be part of an interesting local building, he asked to be taken to the spot where the object had been found. The owner of the hacienda took Julio Tello to the foot of a nearby hill where he saw several similarly marked stones protruding from the ground. Feeling certain that he had stumbled on something important, Julio Tello immediately initiated an excavation. He stayed in Sechin for the next two months, during which time he unearthed the remains of a remarkable monument.

Some 3,000 years ago the people of Cerro Sechin, as the site is now known, had constructed a square building, the front and side walls of which display a line of marching warriors graphically interspersed with images of severed heads, disgorged eyes, eviscerated torsos, infants cleaved in two, severed arms and piles of vertebrae. The message is clear: it is a display of military strength, an unmistakable threat of aggression.

Although Cerro Sechin is unique in early Andean architecture, similar warlike images are to be found on buildings and monuments of Meso-american and Old World civilizations from that period. Moche pottery from Peru also depicts triumphant soldiers celebrating over abject prisoners. This 'iconography of power,' as American anthropologist Joyce Marcus describes it, becomes a common and cogent theme among emerging civilizations throughout the world. As cities evolved into city-states, and city-states into nations, their rulers increasingly indulged in threats and displays of military power. Surely, some people argue, this recurrent aspect of history indicates something about the nature of humanity: that humans are innately aggressive, that war is in our genes.

It is an inescapable fact, as Marshall Sahlins points out, that 'war increases in intensity, bloodiness and duration . . . through the evolution of culture, reaching its culmination in modern civilization.' Human history may be measured in many ways, but one of the grimmest is in the mounting death toll through ever more keenly fought wars. Sigmund Freud said: 'Men are not gentle, friendly creatures wishing for love. . . . A powerful measure of desire for aggression has to be reckoned with as part of their instinctual endowments. . . . Anyone who calls to mind the atrocities of the early migrations, of the invasions by the Huns or by the so-called Mongols under Genghis Khan and Tamerlane, of the sack of Jerusalem by the pious crusaders, even indeed the horrors of the last world war, will have to bow his head before the truth of this view of man.' Freud believed that humans

Part of the carved stone frieze at Cerro Sechin, showing the body of a conquered warrior cut in half at the waist.

are endowed with a basic *instinct* for aggression, and that this is repeatedly manifested on the battlefield.

The meaning of Makapansgat

The discoveries of hominid fossils have often been invoked to strengthen such ideas. The discoveries generally referred to were those from the South African caves, particularly Makapansgat. The *Australopithecus africanus* remains here were intermingled with large numbers of baboon skulls and countless thousands of other animal bone fragments.

In a remarkable series of thirty-nine scientific papers published between 1949 and 1965, Raymond Dart reviewed the evidence from Makapansgat. He saw in the ancient bones of this cave clear signs of the key behaviour that separated us from our ape-like ancestors: hominids became carnivores, he said. He analysed the cause of death of fifty-eight baboons whose skulls were found in the caves, and claimed that depressed fractures in many of the skulls indicated that they had been struck by some kind of bludgeon wielded mainly by right-handed attackers. Raymond Dart outlined his ideas in the following terms: 'On this thesis man's predecessors differed from living apes in being confirmed killers: carnivorous creatures that seized living quarries by violence, battered them to death, tore apart their broken bodies, dismembered them limb from limb, slaking their ravenous thirst with the hot blood of victims and greedily devouring livid writhing flesh.'

When Raymond Dart examined in detail the fossil remains of the australopithecines in the cave, he noticed that they too apparently bore the marks of violence. Our ancestors' killing instincts were not confined to baboons and other animals it seemed, in spite of the fact that at this point in human history hominids still possessed relatively small brains. 'This microcephalic mental equipment was demonstrably more than adequate for their crude, omnivorous, cannibalistic, bone-club wielding, jaw-bone cleaving, Samsonian phase of human emergence. . . . The loathsome cruelty of mankind to man forms one of his inescapable, characteristic and differentiative features; it is explicable only in terms of his cannibalistic origins.'

Summing up his view of human origins in his 1953 essay *The predatory transition from ape to man*, Raymond Dart had this to say: 'The blood-bespattered, slaughter-glutted archives of human history from the earliest Egyptian and Sumerian records to the most recent atrocities of the Second World War accord with early universal cannibalism, with animal and human sacrificial practices or their substitutes in formalized religions and with the worldwide scalping, head-hunting, body-mutilating and necrophiliac practices of mankind in proclaiming this bloodlust differentiator, this predacious habit, this mark of Cain that separates man dietetically from his anthropoidal relatives and allies him rather with the deadliest of Carnivora.'

Raymond Dart's conclusions were radical and his descriptions graphic. In the tense and doom-laden atmosphere of the Second World War and its aftermath, he tried to explain the state of the world in terms of a certain view of prehistory. He later found a sympathetic and eager disciple in Robert Ardrey, playwright, journalist and author. Employing vivid and evocative prose, Robert Ardrey sketched an arresting picture of our past. In a series of popular books, *African Genesis*, *Territorial Imperative*, *Hunting Hypothesis* and *Social Contract*, he described man's supposed predatory, carnivorous and cannibalistic history, and he promulgated the notion that men cannot escape the aggressive instincts that shaped human history. 'Man is a predator whose instinct is to kill with a weapon,' he said in 1961, a view

Makapansgat cave, South Africa. Bones of gracile australopithecines were found here, together with thousands of bone fragments and a large number of baboon skulls. Raymond Dart concluded that the hominids had made a variety of weapons from the bones and used them to kill baboons and other prey animals, and even to murder each other. The results of modern research suggest other explanations for the bone accumulations of these caves.

Above: A reconstruction of the Makapansgat australopithecines with their supposed 'osteodontokeratic' culture, that appeared in 1959.

he still adhered to when he died in 1980. 'We enjoy violence, in our sport and in our entertainment; it's a leftover from our hunting past.'

Robert Ardrey's was not the only voice publicly proclaiming the mark of Cain, and many psychologists and biologists espoused the notion, notably Nobel Prize winner Konrad Lorenz in his book *On Aggression*. Film-makers and writers of fiction pushed the idea yet further into the public domain with works such as *West Side Story, 2001: A Space Odyssey* and *Lord of the Flies*. It was a popular theme, and readily received: humans *are* innately inhumane, and this explains much of the misery, suffering and acts of war in the world.

I believe these ideas are mistaken. In this chapter I will analyse the validity of the evidence for them, examine other aspects of human pre-history that may throw some light on 'human nature' and look at the pattern of urban development since the Agricultural Revolution, a phenomenon that I believe reveals much about the way modern society operates.

Firstly, let us look at the evidence of the bones from the Makapansgat cave. Much of the roof of the cave is gone, removed by the slow process of erosion to reveal a breccia that is astonishingly rich in fragments of fossil bones. So far more than a quarter of a million fossils have been recovered from the site. In common with the other South African caves, it was the discovery of fossilized baboon skulls that suggested the possible presence of hominid remains. James Kitching discovered the first hominid from the site in 1947: Raymond Dart at first named it *Australopithecus prometheus* because it appeared that the creatures had been using fire, but the name was later changed to *Australopithecus africanus* when the inference of fire proved to be mistaken.

Raymond Dart's speculations about the Makapansgat australopithecines fell into two parts. First he interpreted the high incidence of damage in the hominid remains as indicating a remarkable degree of interpersonal violence. Second, after examining in detail some 7,159 fossilized animal bone frag-

Below: Some of the finds from Makapansgat. Raymond Dart suggested that these halves of lower jaws would have been used as saws by the hominids.

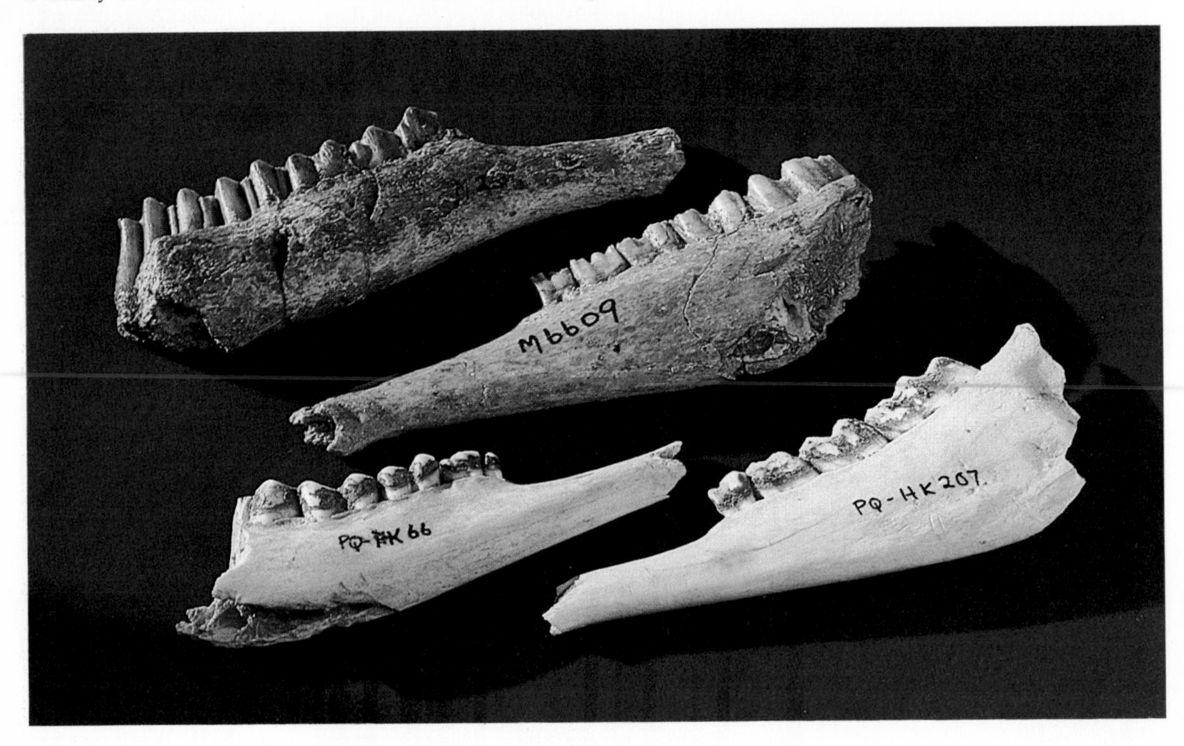

ments from the breccia he concluded that the hominids had brought many of the bones into the cave so as to create weapons and tools from them. Raymond Dart was impressed by the fact that bones from some parts of the skeleton were more common in the deposits than others: surely this indicated that the hominids were selecting the bones that were useful to them? In 1957 he wrote: 'The disappearance of the tails was probably due to their use as signals and whips in hunting outside the cavern. Caudal and other vertebrae may also have disappeared because of the potential value of their bodies as projectiles and of their processes (when present) as levers and points. The femora and tibiae [leg bones] are the commonest of the long bones; probably because they would be most convenient clubs for the womenfolk and children at home.' Part of the upper arm bone (the humerus) he believed to be favoured as a club. Lower jaws snapped in half at the chin served as saws, the double tooth row on a small animal's upper jaw made a useful scraper, while an upturned severed skull formed an excellent bowl. Raymond Dart called this an 'osteodontokeratic' (bone-tooth-and-skin) culture.

A new look at the evidence

Bob Brain, in his work at the Transvaal Museum in Pretoria, is the person principally responsible for encouraging a re-evaluation of Dart's conclusions. For the past fifteen years Bob has been puzzling over the nature of cave deposits. The questions that have mainly concerned him are: how do bones arrive in a cave, and what happens to them once they are there?

Let us look at the second question first. As Bob Brain explains, 'A fossil buried close to the bottom of a deep cave filling may have as much as 30 metres [100 feet] of sediment above it, the weight of which is considerable. The effect of this is to lead to an overall flattening, rather as if the skull had been run over by a steamroller . . . This is why so many of the South African cave hominid fossils are so very distorted. If, however, the cranium

The gracile australopithecine from Sterkfontein cave, known as 'Mrs Ples'. This specimen is rare among the skulls from South African caves, in that it survived intact and undistorted. This happened only because it rolled under a low rock ledge soon after it fell into the cave, and thus avoided being crushed by the weight of the rock layers which later formed above.

becomes filled with matrix, thus giving the brain case internal support, the skull will resist distortion.' But the process can be complicated if a hard pointed object lies close to a skull as it is buried. 'The pressure on the stone can push the point into the skull, causing a localized compressed fracture, just as if the individual had been killed by a blow over the head.'

In Raymond Dart's interpretations of damaged hominid bones one individual was said to have suffered 'a severe transverse blow with a bludgeon on the vertex and tearing apart of the front and back halves of the broken skull'. Of the lower jaw of a twelve-year-old child, Dart inferred: 'The fractures exhibited by the mandible [lower jaw] show that the violence, which probably occurred in fatal combat, was a localized crushing impact received by the face slightly to the left of the midline in the incisor region, and administered presumably by a bludgeon. The result of that decisive blow, as far as the mandible is concerned, was that the four permanent incisors (and perhaps the left second deciduous molar) were sprung from their sockets and the bone was shattered.' Another 'victim' had been killed, according to Raymond Dart, by 'a vertical blow just behind and to the right of the bregma with a double-headed object'. And the Taung child had been attacked with 'a lateral blow on the left fronto-temporal region of the skull'.

Bob Brain has this to say: 'Having observed pressure effects on many hundreds of fossils, I am now extremely cautious when attempting to isolate instances of pre-fossilization bone injury. . . . The damage to the adolescent jaw, for instance, could just as easily have been inflicted by localized pressure from a stone in the deposits. And it's not necessary to assume that the incisors were "sprung from their sockets" by a blow from a bludgeon. These straight-rooted teeth are frequently lost from mandibles prior to fossilization.'

'The evidence is ambiguous,' Bob says. 'Further specimens are urgently needed from situations where ante-mortem damage may be isolated conclusively from post-fossilization effects. Until such evidence is forthcoming, the question of the incidence of interpersonal violence among australopithecines must remain an open one.'

What, then, of the australopithecines' supposed 'osteodontokeratic' culture? The animal bones in the Makapansgat cave *do* represent different parts of the skeleton in unnatural proportions, and the many edges, points, shafts and so on in the bone assemblage could be employed effectively as implements. But what does it mean? Bob Brain comments on bone deposits in this way: 'Occasionally, to the delight of palaeontologists, the entire skeleton of a long dead animal is preserved, with every bone in place and none of them missing. Such an event occurs only in special circumstances, however. . . . More usually, the animal's body is subjected to the destructive influences which characterize any natural environment. The skeleton becomes disarticulated – broken down into its individual parts, each of which has to contend with the attention of carnivores and with the forces of decay and destruction.'

By good fortune, Bob was able to examine at first hand the fate of bones that were once part of a meal. While working at the Namib Desert Research Station, he came across the perfect 'natural experiment'. 'I happened to visit several Hottentot villages and was struck by the number of goat bone fragments which lay about among the huts. Making a small collection of these, I laid them out at the research station and sorted them into skeletal parts as an exercise in osteology. It was immediately obvious that certain parts were well represented, while others were rare or absent. . . . The explanation was not difficult to find – the sample represented the resistant residue of goat skeletons able to survive the treatment which they had received.'

Goat was virtually the only meat available to the village people, and they made use of all that it offered. The leftovers were thrown to the village dogs, who gnawed and licked them until nothing more could be extracted. Many of the more delicate bones were destroyed, only the most tough and resistant parts of the skeleton surviving. Among the collection of bones Bob Brain found that lower jaws were most common, followed by the elbow-end of the upper front limb (the humerus), the knee-end of the hindleg, and the foot bones. Tail bones, the shoulder-end of the humerus, the vertebrae, and the top end of the hindleg bone were virtually absent. Although the villagers had inflicted a surprising amount of damage on the goat bones initially, it was the dogs' teeth that had really stamped the carnivore pattern on the remains. When Bob compared the bone assemblage from the village with the fossil collection from Makapansgat, the result was striking. Although not identical, 'the overall similarity in composition of the bone collections is remarkable,' he says. It would appear that the Makapansgat fossils are not the tools and weapons of an ancient culture, but the leftovers of many carnivore meals.

Evidence to back up this assertion comes from Judy Maguire at the Bernard Price Institute in Johannesburg. Her researches led her into the den of a striped hyaena on the Giora Ilani Nature Reserve in Israel, where she made a list of the types of damage that hyaenas inflict on the bones they take back to their dens. She catalogued nine distinct forms of damage, including pitting, crescent-shaped fractures, random grooves, perforations, splintering and scooping. The comparison with the Makapansgat fossils is illuminating. All nine types of marking are to be found in the fossil collection and, on a very conservative estimate, thirty per cent of the fossil bones bear unequivocal signs of hyaena activity. The incidence of marked bones in modern hyaena dens is close to sixty-five per cent, but the lower figure for the fossil assemblage can be explained, Judy Maguire suggests, by the flaking of the surface in many fossils, which would remove any marks. Although some of her colleagues at the Bernard Price Institute, particularly James Kitching, would disagree, Judy Maguire now concludes that carnivores played a more substantial role in the bone accumulations at Makapansgat than was previously believed.

It is now apparent that much of the information upon which Raymond Dart and Robert Ardrey based their graphic descriptions of our supposedly bloody past simply does not stand up to modern scientific enquiry. In reviewing this work I do not mean to decry Raymond Dart's efforts and major contributions to our knowledge of hominid evolution. As Bob Brain points out: 'Dart's study of the Makapansgat fossils was a pioneering project in that it represented the first analysis and interpretation of a bone assemblage from an African cave.' Other workers had often concentrated on selected samples of bones, looking only at those in which they were particularly interested. His approach marked the beginning of a more logical study of the whole process by which bones become preserved and fossilized. Indeed, it was Raymond Dart who in large measure encouraged both Bob Brain and Judy Maguire in their efforts to test his earlier hypothesis. Ironically, he has been much more ready to accept the implications of their findings than have many of his supporters.

The effects of settling down on the !Kung

Most hunter-gatherers settled down and became agriculturalists many millennia ago. A few hunter-gatherers still exist today, but many of these

The entrance to a hyaena den in Amboseli National Park, Kenya. Raymond Dart considered the suggestion that the Makapansgat bones might have been accumulated by hyaenas and a few hyaena dens were looked at, but none of them contained large numbers of bones. More recent studies have shown that hyaenas are rather variable in their habits and that some do collect piles of bones at their dens. Furthermore, as at Makapansgat, some bones are far more common than others, simply because they are more durable and survive the hyaenas' crunching jaws. Telltale marks are left on the bones by the hyaenas' teeth and comparable marks are found on a number of the Makapansgat specimens.

groups are currently in the process of adopting a sedentary existence, and this offers an important opportunity for studying some of the social implications of the change. Some of the !Kung are now undergoing such a transition, and they are being closely studied. 'There is a major contradiction in the transition through which the !Kung are now passing,' says Richard Lee, 'and this is between *sharing*, which is central to the hunter-gatherer way of life, and *saving*, or the husbandry of resources, which is equally central to the farming and herding way of life. The food of the !Kung camp is shared out immediately with residents and visitors alike; for herders to do the same with their livestock, or farmers with their grain, would quickly put them out of business. Many families are suffering because of these conflicting demands.'

Richard Lee tells of one !Kung man, named Debe, who appeared to be settling into the new way of life very well. He had assembled a small herd of goats and cattle and was becoming a successful herder. But when meat was scarce among his hunter-gatherer relations at nearby /Xai/Xai, they visited and asked to be fed. 'Under heavy social pressure Debe would slaughter one goat after another, until after several years he sold or gave away his remaining herd saying that the responsibilities were too heavy,' Richard Lee explains.

Another man, Bo, successfully raised six cows and nurtured fields of maize and melons. Knowing the danger of succumbing to continual requests for food from his relatives, he would welcome visitors, but send them on their way after just one meal. 'The effect of this behaviour was striking,' says Richard Lee. 'People spoke of Bo as stingy and far-hearted; he became feared, and there were mutterings that he had learned techniques of witchcraft from black medicine men. Then his son's wife left her husband and other kin shunned Bo's camp leaving Bo a successful but isolated farmer. Finally, in 1970, Bo had had enough. He sold his cattle and other stock for cash, packed his things, and walked across the border to settle at Chum !kwe in Namibia.' There are many such examples that underline the sharp distinction between the farming and the hunter-gatherer ethic.

The principal reason for the !Kung's gradual adoption of agriculture is pressure from the government. The !Kung have for years lived in close contact with agriculturalists whose herding and agricultural practices they are now to some extent copying. A typical !Kung field is less than 1 hectare (2.5 acres) in size, roughly oval in shape, and usually has no irrigation. Their most important crops are maize, melons, sorghum and tobacco. Tobacco is by far the most difficult of the crops, requiring constant watering and good shade. 'The fact that the !Kung devote so much time and effort to tobacco, a non-food crop, suggests that the motive of increasing food supply is not uppermost in their minds,' Richard Lee observes.

Most of the work in tending fields falls to the women. The men spend a lot of time with their animals, so that they are away from the village to a much greater extent than their wives. 'This division of labour is turning out to have important political implications,' says Richard Lee. 'For a start, men's work with the herds brings them into contact with the local agriculturalists, an opportunity that is mostly denied to the women. And as the local political structures are becoming more and more important to the !Kung, this means that, unlike previously, power is very much a prerogative of the men. Secondly, livestock are often sold when animals are mature, and this gives the men access to a cash economy. The women's agricultural produce is mostly for subsistence, so once again they are more or less barred from an important new institution. They do make a kind of honey beer which they sell, but this is not significant compared with the income from livestock.'

Patricia Draper has been making a special study of social interactions in the !Kung who are changing their way of life and she says that her 'strong impression is that the sexual egalitarianism of the bush is being undermined in the sedentary !Kung villages'. Sex roles are becoming more rigidly defined, she claims, with men now being inclined to shun some jobs as 'unmanly' or 'unworthy' of them. This was rare previously. 'Women in bush life derive self-esteem from the regular daily contribution they make to the family's food. . . . They also retain control over the food they have gathered after the return to the village,' Patricia Draper reports. The changed status of women in the new way of life is striking.

A !Kung camp in the bush is a tightly knit collection of seven or more simple shelters arranged in a circle facing the centre. Patricia Draper describes the life there in the following way: 'Everyone in the camp can see (and often hear) everyone else virtually all the time, since there are no private places to which people can retire. Even at nightfall people remain in the visually open space, sleeping singly or with other family members around the fires located outside the huts.' Intimacy is the over-riding characteristic of such a settlement. Gossip is exchanged, disputes are resolved, food is shared – everything happens within this intensely social focus.

Life in the farming village is different. Here, the houses are dispersed, often around a cattle corral. The house entrances no longer directly face the life of the village. People are no longer in close physical contact with each other. 'With the flow of food in a nomadic settlement goes a flow of emotions and feelings,' claims Richard Lee. 'That flow is stemmed in the farming village.' A combination of a settled life and access to cash has also led people to begin to accumulate material wealth in the form of clothes and other 'consumer goods'.

Some of the !Kung are adapting very readily to the new way of life while others prefer the old way. Patricia Draper asked some !Kung women which was the better life, that of a !Kung woman or a Herero woman. Kxarun !a, a woman of about fifty, replied: 'The !Kung women are better off. Among the Herero if a man is angry with his wife he can put her in their house, bolt the door and beat her. No one can get in to separate them. They only hear her screams. When we !Kung fight, other people get in between.' Her companions agreed earnestly with her. Clearly, for these women there is more to life than may be offered by agriculture and the material progress that comes with it: their intimate social structure *is* their life.

The shift to agriculture is having its effect on !Kung children too. 'In the bush, children of both sexes do equally little work,' explains Patricia Draper, 'and their play isn't differentiated by sex, boys doing one thing and girls another. . . . In settled villages, however, there is a shift towards viewing a child as a potential worker. Boys, for example, help with herding the animals, while girls may do some work with their mothers in the fields or help in the house.' Patricia Draper has also noticed a distinct development of 'girls' games' and 'boys' games', with the more sharply defined male and female roles in adults apparently being echoed in the maturing young.

One very practical consequence of the !Kung's new way of life is a marked rise in the birth rate. Whereas women previously had babies once every four years, the rate in recently settled agricultural villages is closer to one every two-and-a-half years. And in one community that made the change to agriculture early on in the 1960s, the births are now separated by less than two years. Another change is a growth in possessiveness over the land from which they make their living. Nomadic hunter-gatherers move camp frequently, so as to exploit food available in different areas, and parts of a band may pack up and depart from the rest to resolve tensions between group members. Nomads are free to move as they wish for their foodstore is in every part of the land. This is not true for farmers. Clearing ground, sowing seeds, nurturing the young seedlings, and harvesting the crop all demand that a farmer stays put. 'When people's livelihood is rooted in the fields, they are embedded in the land in a way that hunter-gatherers are not,' comments Richard Lee.

This point is certainly crucial in understanding what is perhaps the primary implication of agricultural food production as against nomadic food gathering: as soon as people commit themselves to agriculture, they commit themselves to defending the land they farm. To run away in the face of hostility is to face certain loss: a year's labour may be invested in the fields, and that cannot be given up easily.

As well as land that requires defending, agriculturalists tend to acquire property, both personal and communal, that needs to be guarded. Even without the evidence of human history, one might expect that a substantial increase in military encounters between neighbouring groups would have followed the Agricultural Revolution. In the final part of this chapter I will look briefly at some early examples of territorial aggression in the hope that this may help put into perspective the politics of the twentieth century.

A Babylonian boundary stone recording the purchase of land by a government official. Hunter-gatherers move about freely, collecting whatever food the land has to offer, but farmers develop a far more possessive attitude towards the land in which their livelihood is rooted. Only with agriculture do areas of land become private property, to be bought and sold, or even fought over.

The growth of the early cities

It is in the origin and growth of cities that one sees the blueprint of the future of mankind. Villages and small towns were the immediate products of the Agricultural Revolution, but these were at least based on the social web of the nomadic band. With the city, however, that web was spun into a much more complex and clearly defined fabric. As Richard Lee puts it, 'You can't organize 500 people in the same way as 50, still less 5,000. Social, economic and political institutions have to be constructed. You get chiefs, arbitrators, elites, and probably formalized religion.' In short, the city depends upon closely organized activities that respond to control from centralized authorities. The craftsman of former village or town life becomes part of a larger venture in which he might be required to devote some of his work to corporate ends. In return he receives the social and economic benefits of city life and protection by an army.

One of the true marks of 'city' status is the community's ability to engage in constructing public works such as temples, palaces and canals. With the city of Jericho, it was the famous wall. This ancient city, which may once have housed 3,000 people in its 4,000 square metres (10 acres) lies in a rich green oasis just north of the modern city of Jericho in the Jordan Valley. The earliest inhabitants of 10,000 years ago built simple round houses from sun-dried mud bricks and they made no pottery. But they combined their efforts in organized and controlled labour to erect a wall of impressive proportions: some of it still stands today, reaching up to 4 metres (130 feet).

The old town of Jericho was occupied periodically for almost 7,000 years. At least ten peoples of distinctly identifiable cultures lived there over that time, exploiting the rich farming potential of the arid but otherwise fertile soil. The city's prime position on the long-established north–south trade route brought commerce to the city and its merchants: obsidian, greenstone, hematite, sea shells and salt all passed through the city's trading institutions.

What is one to make of the wall of Jericho, besides the fact that the people of the city were ruled by a central authority strong enough to plan and execute public works? It may have been built for military defence, but it is just as likely that the wall was erected to keep out silt, rather than people. Situated as it is near the slopes of mountains, large quantities of silt were probably washed down the valley during floods and could have buried the town's buildings. Many people today believe that the wall of Jericho, far from being the earliest piece of evidence for organized aggression was simply a protection against the effects of floods.

The rise and fall of Monte Albán

Some recent discoveries in the New World illustrate clearly some of the processes through which civilization has passed. One is in the Oaxaca Valley in the southern highlands of Mexico and the other is in coastal Peru.

The Valley of Oaxaca was the location for one of the earliest urban developments in Mesoamerica: the city-state of Monte Albán. The city had a relatively brief but unquestionably powerful history, beginning around 2,500 years ago, rising to a peak of influence some 600 years later, and then falling into decline about 1,300 years ago. During this period, Monte Albán's rulers first of all flexed their military muscle in order to achieve domination over neighbouring territories and then devoted a good deal of time and effort to establishing and maintaining diplomatic relations with nearby centres of power, such as the Aztec city of Teotihuacán to the north

The remains of the tower of the ancient city of Jericho. Like the city wall, it was built about 10,000 years ago.

Mesoamerica. The small map shows the areas occupied by the main tribes, and the larger map shows the location of the Oaxaca and Tehuacan Valleys. Rapid population growth in the Oaxaca Valley followed the adoption of maize as a staple crop about 3,000 years ago. Within 1,000 years social structures had changed radically, and many people from the scattered villages were drawn into a massive urban centre, Monte Albán. This city gained control of a wide area, but its rulers maintained peaceful diplomatic relations with the Aztec empire centred on Teotihuacán.

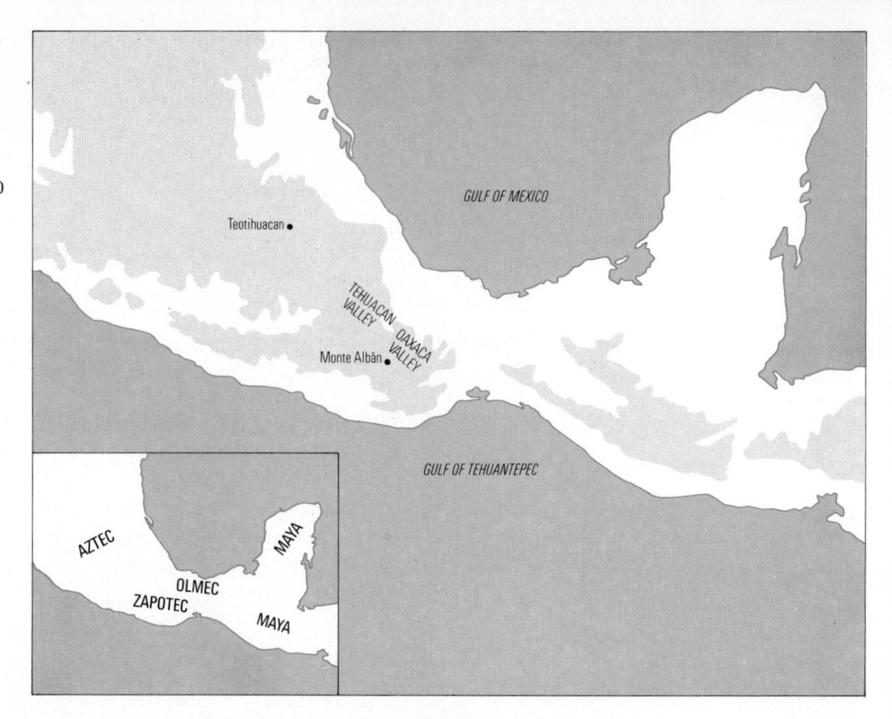

and the Gulf states to the east. The urban centres of these major political groups all declined at about the same time.

The previous chapter described the pre-agricultural history of Oaxaca and told how it 'took off' as soon as there was a commitment to maize cultivation. Before Monte Albán was established the Valley of Oaxaca was occupied by a handful of competitive chiefdoms, each centred on a large village with civic or ceremonial structures, and surrounded by smaller hamlets. About 2,500 years ago the rival groups appeared to collect themselves together as some kind of Zapotec confederacy. As the centre of that confederacy they founded the city of Monte Albán on a 400-metre (1,300-foot) high hill close to the confluence of two rivers, the Rio Atoyac and the Rio Salado.

Monte Albán's inhabitants soon numbered 10,000, settlers being drawn from neighbouring communities under its rule. The population doubled within a few hundred years, and peaked at possibly 30,000 within 700 years of its foundation. At this point the city covered approximately 6 hectares (15 acres), and it governed another 30,000 people living in the 700 square kilometres (270 square miles) that formed its political sphere of influence. The collapse of the city in AD 700 was marked by a massive population decline, leaving a residue of perhaps 6,000 inhabitants.

The vanished city population does not necessarily imply some catastrophe or military holocaust: indeed the eclipse of Monte Albán seems to have been marked by peaceful decay rather than a cataclysmic end. The city dwellers probably migrated back to the countryside, re-establishing the neglected villages and towns. Whatever it was that had drawn the people together in the first place, and had held them firmly in a powerful urban concentration, no longer applied. This phenomenon of city growth through recruitment of local inhabitants, followed eventually by a dispersal back to the rural villages, is common in the early history of urbanization: it is seen frequently in the Old World centres of the Sumerian civilization, for instance.

Clues to the character of Monte Albán's political evolution come from the nature of its buildings and decorations. One of the first major public works undertaken by the people of the city, as at Jericho, was the construction of a substantial wall, about 3 kilometres (2 miles) long and almost four metres (13 feet) high in many places. Rising as a prominent building in the city soon after the wall was built was, according to Joyce Marcus, 'one of the most impressive works of military propaganda in all Mesoamerica'. Along the east face of the building was a gallery of 300 or more grotesque figures carved into the stonework. Naked bodies of what were, presumably, prisoners are displayed mutilated and bleeding. Whatever the precise intent of the gallery, it was certainly not constructed in the name of peace.

'This is the kind of propaganda one associates with an emerging state that is fighting to take control over previously autonomous regions and wants to discourage resistance,' explains Joyce Marcus. 'Of the shared conventions in Mesoamerican iconography some of the most widespread are those depicting captives. Prisoners are displayed in humiliation; they are stripped naked and bound, and their posture is awkward. The captors, in contrast, are dressed in elegant regalia and are posed in rigid dignity. If a prisoner has been sacrificed, he is shown with his eyes closed and his mouth open, and in many instances with flowery scrolls, presumably representing blood, issuing from his wounds.' In many Mayan ceremonial buildings, for example, images of abject prisoners were carved into stone staircases, so

One of the oldest buildings in Monte Albán, erected shortly after the wall was built. Along its east face were placed a series of carved stone slabs showing over 300 figures of naked and mutilated prisoners. One such slab, which has since been removed from the building, can be seen propped up in the foreground.

The central plaza of Monte Albán, caught in the rays of the rising sun. During the height of its power, as many as 30,000 people lived in this city.

that the victors could literally walk over the bodies of the vanquished. 'The Aztec built displays that served a similar purpose: the *tzompantli*, a rack or wall consisting of skulls of enemy dead,' Joyce Marcus adds.

The rulers of Monte Albán clearly felt it necessary to underline their military power by erecting such monuments in the city. Indeed, monument building during this early period outstripped anything that followed. An arrow-shaped building was constructed relatively early in the history of Monte Albán at the south end of the city's massive central plaza. It appears to chronicle victories in a series of forty so-called 'conquest slabs' set into the walls. A combination of hieroglyphs represent place names of a conquered locality, together with the style of dress worn by the elite of the area.

Monte Albán's expansionism continued successfully for several centuries, involving displays of terror tactics outside the city as well as within its walls. At one settlement, for instance, the military followed the Aztec practice of building a skull rack, and strong fortifications were clearly necessary at the limits of the province. But about 1,900 years ago, the growth of the city-state was checked, apparently because it ran up against a

similarly expansionist Aztec state based at Teotihuacán, 500 kilometres (300 miles) to the north, in the basin of Mexico.

The two 'superpowers' apparently had a good deal of respect for each other, for there is no sign in the Monte Albán iconography of any military conflict with the Teotihuacanos. Instead, there are many records of ambassadorial meetings on the buildings of this later period. Meetings with the people of the Gulf coast were accompanied by the exchange of exotic gifts. For example, obsidian and magnetite, a shiny mineral used in decorative objects, were taken from the highlands to the coast, as were the beautiful green and blue feathers of macaw parrots. In exchange, the elite of Monte Albán received turtle-shell drums, armadillo shell and crocodile teeth. Gift objects similarly flowed between the Zapotecs and the Aztecs. The rulers of the regions had apparently abandoned their obsession with warfare and were developing the art of diplomacy.

The city of Sechin Alto

The gallery of slain prisoners built in the early stages of Monte Albán's history is clearly very reminiscent of the images at Cerro Sechin, the 3,000-year-old monument from coastal Peru with which this chapter opened. Joyce Marcus is struck by the similarity and sees Cerro Sechin as representing a stage of urban development akin to that in the Zapotec city. Lorenzo Samaniego, the archaeologist who in recent times has been most responsible for the excavation of Cerro Sechin, suggests that it might have been erected specifically to commemorate a victory of the local people over the inhabitants of the neighbouring valley of Moxeke. Certainly, at this point in the history of Peru's coastal communities, there were many rival chiefdoms existing uneasily side by side, and they may have buried their differences and formed a confederacy, just as happened in the Valley of Oaxaca. Large urban centres with a similar role to that of Monte Albán were probably built as a consequence. One such urban centre was Sechin Alto, a large settlement contemporary with and just a few kilometres distant from Cerro Sechin. Sechin Alto was discovered by Julio Tello while he was excavating Cerro Sechin.

So far this larger site, which covers something in the region of 5 to 10 square kilometres (2 to 4 square miles), has not been excavated. Nevertheless, it is clear that buried there are the remains of a massive, thriving community. Probably founded by marine-based societies who moved inland 3,000 years or so ago, Sechin Alto was principally an agricultural community, with maize its staple crop. The long sweeping Sechin and Casma valleys were once extensively irrigated under the central authority of established rulers at Sechin Alto. Military power probably had to be exercised to make the local inhabitants participate in this venture and the stark images of Cerro Sechin are doubtless a powerful reminder of that.

Sechin Alto grew steadily but strongly over a period of centuries, the main mound being built outwards and upwards in a deliberate effort to establish and represent the power of the central authority. (Although the mound has not been excavated, it is obvious, from the way it is now disintegrating, that it was built up in successive stages, rather than all at once.) The mound was as elegant as anything in South America, with terraces, walls, plazas and private rooms decorated with plaster relief. Shaped overall like a giant's armchair, the mound faces east across the valley in the direction of deliberate clearings in the foothills 5 kilometres (3 miles) away. A total of at least six million tonnes of rock and adobe brick went into the construction of this monumental project, which implies public work on a grand scale.

Standing on the summit of the main mound at Sechin Alto, one sees smaller mounds to the right and left, and a series of three circular plazas, each 80 metres (250 feet) in diameter, stretching out processionally in a line towards the east. Sadly, one of them now has the Pan American highway going through it. Remains of terraces, walls and other major buildings mark out this ancient mall, reaching eventually to the clearings in the distant foothills that presumably acted as orientation markers for laying out the long mall. Directly behind the main mound are three small 'giant's armchairs' joined together, and to the northwest is a much smaller collection of structures that, in miniature, echo Sechin Alto proper. The whole site is indeed extremely impressive.

One can imagine the site I have just described as being the political centre of a large community, the architectural expression of a highly structured society. No domestic buildings remain, not even as buried foundations, for these would be apparent on the aerial photographs that have been taken of the site. The houses were probably of reed and adobe

The enigmatic ruins of Chancayillo. This strange construction, with its three concentric walls and baffled entrances, may have been a fort, although the tiny rooms inside its walls could not have housed an army of any size.

and would have been washed away in the occasional devastating floods. Andean civilizations for a very long time operated a system of labour tax: people had to work part of the time for the state, either directly in erecting public buildings or indirectly by setting aside a proportion of their agricultural or other produce. There are records of this in the Inca civilization, where it was called the *mit'a*. Builders' marks on the adobe bricks in the Moche Huaca del Sol, probably the biggest man-made structure in South America, illustrate the work of separate labour units as do the walls of Chan Chan, the Chimu capital. It would seem likely that Sechin Alto is also the product of labour organized in this manner, although as yet there is no evidence for this.

Looking southeast up the Sechin valley one can just make out a large structure in the distance, a building perched on the top of the valley side. This construction, called Chancayillo, is an enigma. It dates from the time of Cerro Sechin and Sechin Alto and was undoubtedly built by people working under the labour-tax system. The most obvious interpretation is that Chancayillo was a spectacular fortification, guarding the irrigated fields in the valley below, but this may be wrong. Chancayillo has three concentric walls forming the outer defences (if indeed that was the function of the structure), each one breached by a small 'baffled' entrance. The diameter of the outermost wall is 150 metres (500 feet). Inside are several double-walled buildings, again with the protection of baffled entrances. Indeed, the extent of the multiple walling and the generous door baffles means that the internal rooms are extremely small. The whole 'fort' has very little room in which to house a substantial militia, and Chancayillo could possibly be some form of temple, but the immediate impression it gives is of a fortification. If this is correct, then it is yet another indication of the rise of warfare that accompanied the early days of sedentary agriculture and city formation.

The aim of this chapter has been to present warfare as a social and political response to changed economic circumstances. However, it was the nature of *society* that changed with the transition from nomadic hunting-and-gathering to sedentary agriculture, not the nature of man. Humans are essentially cultural creatures capable of responding in many different ways to the same prevailing circumstances. In the next, and final, chapter, I will explore further what is meant by 'human nature' in the light of what can be learned from history and prehistory, and will offer a personal view of what this implies for the future of the human race.

14 The Future

On the cover of the *Bulletin of Atomic Scientists* is the face of a clock. Its hands are set at seven minutes to midnight: 'a symbol of the world's approach to a nuclear doomsday,' explains the Bulletin's editor, Bernard Feld. The editorial in the journal's first issue for 1980 began with these chilling words: 'As the Bulletin begins its 35th year, we feel impelled to record and to emphasize the accelerating drift toward world disaster in almost all realms of social activity.'

As I write the words of this closing chapter, stories of global crises compete with each other for space on the newspapers' front pages. Tension and suspicion mount between East and West while the less newsworthy toll of death and disease continues to rise in the poorer nations of the world. The first shudders of panic can be detected as it is realized that certain important natural resources, particularly oil, are indeed limited, and may soon be exhausted. The global economy is under even greater danger of collapse through oil-fuelled imbalance and political uncertainty. Terrorism strikes viciously and unpredictably, signalling a growing anarchy. As Bernard Feld implied in his editorial of 1 January 1980, never before has mankind faced a future of such insecurity and potential danger.

Prophecies of doom are not new. In every generation there have been some who have seen technological and social progress as leading to the destruction of civilization. Sigmund Freud wrote the following half-a-century ago: 'Men have brought their powers of subduing the forces of nature to such a pitch that by using them they could now very easily exterminate one another to the last man. They know this – hence arises a great part of their current unrest, their dejection, their mood of apprehension.'

Within a decade of these words having been written, the Second World War was being fought, with massive destruction of cities and the loss of millions of lives. The war did not result in the extermination of every last man as Freud had feared, but it ended with a clear signal that this possibility was now very near. That signal was, of course, the annihilation by atomic weapons of Hiroshima and Nagasaki. Hiroshima was the first civilian target for an atomic weapon, and there were close to half-a-million victims from the detonation of a single atomic bomb that, by today's standards, was of very modest power. Modern nuclear arsenals house warheads with a total explosive power of *one million* Hiroshima bombs.

The Third World War would almost certainly be the very last war. The combination of gargantuan destructive power with extraordinarily sophisticated tracking, guidance and delivery systems would ensure that an all-out conflict resulted in irrevocable loss to all parties involved. The rest of the world, though not involved, would be showered by lethal radioactive fallout. Our planet would be completely devastated, and almost

An Indian family living in extreme poverty. Three-quarters of the world's population live in the poorer countries of the southern hemisphere, yet these countries share only a fifth of our planet's wealth. Sadly, the governments of many poor nations spend valuable resources on advanced weapons rather than helping their neediest people. Thus the enormous gulf between rich and poor is only widened by the atmosphere of tension between East and West.

all forms of life, animals, plants and bacteria, would suffer the same fate.

Some of the details of the superpowers' weaponry are puzzling. The United States and Russia between them have 16,000 nuclear warheads, although there are only 400 cities in the northern hemisphere large enough to 'justify' attack by such a weapon. Indeed, a United States Poseidon submarine carries enough nuclear warheads to destroy all Soviet cities with a population greater than 150,000, and yet the United States has thirty-one such submarines. One observer who witnessed an early nuclear test described the sound as being 'like the gates of hell slamming shut'. If some of the weapons that are now ready to be fired at the push of a button were ever in fact used, those gates would indeed slam shut, very firmly, and for good.

The arms race that has brought our world to the very brink of destruction has been stimulated by many factors. The momentum of vast military machines is considerable, but ideological differences and simple suspicion are powerful forces too. The world peace of the past three-and-a-half decades has rested precariously on an ever more uncertain balance of terror. Despite the Strategic Arms Limitation Talks (SALT) of the past ten years or so, the development of weapons has continued apace. The cost of this military adventure in global terms is 450 billion US dollars per year, twenty times as much as the richer countries of the world give in aid to the poorer countries. The energy and determination with which military superiority is pursued surely indicates something about the nature of mankind, but what does it indicate?

Aggression in the animal kingdom

Male sticklebacks react instinctively to the red belly of another male when they are both seeking mates, and a challenge ensues. Vervet monkeys chase intruders from their territory amid much screeching. A red-deer stag will engage antlers with a harem-holder if he thinks he can defeat him and take over the harem. Such acts of 'aggression' are a familiar part of animal life. Some people therefore argue that human beings must also be endowed with aggressive instincts. They suggest that the social conformities of civilized society effectively stifle such impulses for much of the time, but that these aggressive instincts cannot be suppressed indefinitely.

There are two issues here. The first concerns aggression in the animal world and whether this equates with human aggression and warfare. The second is the question of how much human behaviour is, or is not, determined by what is in our genes.

Conflicts over mates, food and territory are commonplace in the natural world, but animals have an elaborate style of combat when settling a dispute with another member of the species, and, for the most part, they avoid inflicting serious damage on each other. One of the important factors about animal conflict is that it is more likely to happen under some types of circumstances than others. Factors such as availability of resources and territorial crowding have an important effect on the level of violence running through a group of animals. Under crowded conditions, for example, aggressive acts are far more likely to happen.

One has to ask, however, whether or not it is valid to observe a nation preparing for war and compare this with, say, a baboon displaying its sharp canine teeth when a rival male appears. It is part of a baboon's nature to react under certain circumstances by giving the characteristic 'yawn-threat'. If a baboon did not occasionally engage in this form of display the animal would find it very difficult to integrate well into its natural community. The animal would not truly be a baboon. What of

A male baboon's 'yawn threat'. This display of the sharp canine teeth is usually sufficient to deter a rival. By means of such signals, social hierarchies are maintained in animal groups. Outright conflict is rare and even where fighting does occur, serious injuries are not generally inflicted.

humans and warfare? Is war an inextricable thread in the fabric of human nature?

The importance of culture

Running through the history of human evolution is a persistent theme, that is, the elaboration of material and social culture. Most animals interact with the world on the basis of instinctive behaviours, modified to some extent by their own experience, but, as David Pilbeam emphasizes, 'Man is a learning animal *par excellence*. We have more to learn, take longer to do it, learn it in a more complex and yet more efficient way (that is, culturally), and have a unique type of communication system, vocal language, to promote our learning.' Humans come into this world equipped with very few instinctive responses: suckling, crying, smiling and walking may well be the only things human beings do instinctively. What a person eventually becomes, in terms of both behaviour and beliefs, depends on the culture in which that individual is immersed.

Man not only makes culture, but is also made by culture. David Pilbeam explains: 'Whether we have one or two spouses, wear black or white to a funeral, live in societies that have kings or lack chiefs entirely is a function not of our genes but of learning.' Clifford Geertz has stated the relationship between humans and culture thus: 'Without men, no culture, certainly; but equally, and more significantly, without culture, no men.' We come into the world with the potential to lead any one of a thousand lifestyles. But we lead one that is shaped by the cultural traditions into which we were born.

The endless variations of cultural styles, including religious beliefs, social rules, styles of dress and language testify to our extreme flexibility.

There are no universal rules which people throughout the world obey. Even the prohibitions on murder and incest, though found in most societies, can be broken in some. *Homo sapiens sapiens* is unquestionably the product of natural selection, but the principal characteristic of our behaviour is that it is moulded by the society in which we live.

Against this background it is ludicrous to argue that organized warfare is equivalent to the baboon's aggressive baring of its canine teeth. National leaders who engineer military conflict with another nation are engaged not in aggression but in politics, and the individuals on the battlefields are more like sheep than wolves. Hand-to-hand killing is no doubt carried out in an atmosphere charged with emotion and anger, but think how much indoctrination and depersonalization has been performed in order to bring combatants to this state of mind.

Speaking from the perspective of a prehistorian, Bernard Campbell has this to say: 'Anthropology teaches us clearly that Man lived at one with nature until, with the beginnings of agriculture, he began to tamper with the ecosystem: an expansion of his population followed. It was not until the development of the temple towns (around 5000 BC) that we find evidence of inflicted death and warfare. This is too recent an event to have had any influence on the evolution of human nature. . . . Man is not programmed to kill and make war, nor even to hunt: his ability to do so is learned from his elders and his peers when his society demands it.' It seems ironic that the capacity for culture, which is shared equally by all the peoples of the world, should be the instrument which also erects barriers between people. Different religious beliefs and ideologies have many times in history been the cause of hatred and conflict.

Most ironic of all, however, is the divisive effect of language. No other creature has the human capacity for spoken language, and it is the foundation on which culture is built. Without language, complex social systems and sophisticated technology would be impossible. It is a capacity that brings people together, but it also divides different groups of people through its principal artifact, culture, and because the great diversity of human languages erects barriers to communication between groups.

I believe that the nature of man is more complex than is usually supposed. We do not carry with us the burden of a more primitive and savage past: humans are not 'killer-apes' as has been suggested. Nor are we innately peaceable creatures. Natural selection has endowed us with a behavioural flexibility which is quite unknown in the animal world. Without doubt we are highly social creatures, and in the absence of other individuals with which to interact we would not be human. For several million years our forebears pursued a way of life, hunting-and-gathering, that demanded a degree of co-operation not displayed by other primates. But it would be as wrong to assert that humans are innately co-operative as it would be to say that we are innately aggressive. We are not innately *anything*. Humans are cultural animals, and each one of us is the product of our own particular cultural environment.

Those who believe that man is innately aggressive are providing a convenient excuse for violence and organized warfare. Still worse, such beliefs increase the likelihood that the holocaust which is predicted will indeed come to pass. It is of little consequence how we envisage the nature of the physical universe: the planets continue in their orbits around the sun whether we believe in them or not. But, as psychiatrist Leon Eisenberg stresses, 'The behaviour of men is not independent of the theories of human nature that men adopt.' When, for instance, the nature of mental illness was thought to be marked with violence, 'patients' were chained, beaten and locked up. They reacted with violent fits of rage, thus fulfilling society's image of them. If we allow ourselves to believe that humans are innately aggressive and that we are inevitably driven towards conflict, then nothing is more certain than the eventual fulfilment of that belief.

The gap between North and South

So far I have concentrated on the military might of the superpowers and the threat this poses to the future of our species, but there are other problems which add to the instability of the world situation. One of these is the massive inequality in the distribution of wealth. A quarter of the world's population lives in the 'developed' nations of the northern hemisphere, and they account for four-fifths of global income. The poorer countries of the southern hemisphere (excluding Australia and New Zealand) are inhabited by the remaining three-quarters of the population who share just one-fifth of the world's wealth. This is the extent of the gulf between North and South.

A little more than a decade ago the United Nations Secretary-General, U Thant, issued the following warning: 'The members of the United Nations have perhaps ten years left in which to subordinate their ancient quarrels and launch a global partnership to curb the arms race, to improve the human environment and to supply the required momentum to developing efforts.' U Thant expressed his fear that the problems might already be too great to be controlled. 1980 saw the publication of a major report by an international commission chaired by Willy Brandt. Entitled *North-South: a programme for survival,* the report underlines the social, economic and health problems that divide the world's rich from the world's poor, and argues that unless the North recognizes its moral responsibility for the South, global disaster is inevitable. The report makes it clear that the problems have, if anything, deepened rather than improved since U Thant made his gloomy announcement in 1969.

'There is a real danger that in the year 2000 a large part of the world's population will still be living in poverty,' wrote Willy Brandt in his introduction to the report. 'The world may become over-populated and will certainly be over-urbanized. Mass starvation and the dangers of destruction may be growing steadily – if a new major war has not already shaken the foundations of what we call world civilization.'

Although the population explosion is abating somewhat, world population currently stands in excess of 4,000 million and will have reached 6,000 million by AD 2000. Of today's 4,000 million at least a quarter are seriously undernourished and ten per cent are actually starving. These people are suffering, not because our planet lacks the required resources, but because the resources we have are unfairly distributed. The terrible truth is that there is enough food produced for everyone in the world to be adequately fed, but food goes not to those who need it most. It goes to those who can pay most for it in quantities that far exceed their needs. Unjust social and economic structures create the poverty that is the cause of starvation and the wealth which allows food and other resources to be wasted in the 'developed' countries of the world.

The *North-South* report also gives some telling examples of how military spending channels funds away from practical problems that affect the lives of the world's poor. For instance, the military expenditure of just half-a-day would finance the whole malaria eradication programme of the World Health Organization. Even less would be necessary to overcome river blindness, a condition that affects millions of people. The cost of just one tank would pay for effective storage facilities for 100,000 tonnes of rice, thus saving at least 4,000 tonnes annually. Forty thousand village

LONDON

A rubbish tip in England illustrates the careless waste of resources in the richer countries of the world. Unjust economic structures, dating from colonial times, create the wealth which allows the northern hemisphere to squander resources, including food, while starvation is widespread in the poorer nations of the southern hemisphere.

pharmacies could be provided for about twenty million US dollars, the price of one jet fighter. The comparatively small sum of twenty-three billion US dollars would pay for the farm equipment required to bring the poor countries close to self-sufficiency in food by 1990. And a mere nine billion US dollars would be sufficient to provide a safe water supply and hygienic waste disposal for all the countries now lacking these facilities. The richer nations, incidentally, spend a hundred billion US dollars every year on alcoholic drinks. The list is long and the comparisons appalling.

The gulf between rich and poor nations is now so wide that the people at either end of the spectrum are living in entirely different worlds. And yet the governments of many poor nations, rather than helping their poorest citizens, are spending their resources on advanced weapons. 'It is a terrible irony,' says Willy Brandt in his introduction to the *North-South* report, 'that the most dynamic and rapid transfer of highly sophisticated equipment and technology from rich to poor countries has been the machinery of death.' There is an inescapable link between the obsession with advancing military capability and the neglect of the world's poor. It is clear that the atmosphere of suspicion between East and West is contributing substantially to the economic disparity between North and South.

Choosing our future

Many observers have pointed out that the human race occupies one world. We share the same planet, they argue, and we should therefore share what it has to offer and avoid spoiling it for others. From the perspective of prehistory, one can also add that we are one species, and that every human being in every part of the globe shares a common heritage with every other. This, I believe, is a powerful motivation for reconsidering the blatant inequities in the world before such imbalances drive us all into oblivion.

The work I have described in this book surely demonstrates that *Homo sapiens sapiens* derives from a single stock and that the physical differences between people in different parts of the world are simply the kind of geographical variation one would expect in a widely distributed species. The differences between people are, in effect, skin deep, and this is an apt metaphor when one considers the long history of social oppression based on skin colour.

The inhabitants of different regions *are* different, there is no denying that. As I said earlier, the most pronounced differences are in the ways people do things: their dress, their architecture, their myths, their songs, their ideals and so on. This should be seen as a manifestation of the enormous inventiveness of human culture, not as a cause of division. The earth is populated by one people living many different styles of life because of a unique cultural capacity. And the mind that expresses this unique capacity is the one that also universally seeks beyond itself for explanations of man himself and the nature of the world around him.

'Among the two million or more species now living on earth,' wrote Theodosius Dobzhansky, 'man is the only one who experiences the "ultimate concern". Man needs a faith, a hope and a purpose to live by and to give meaning and dignity to his existence.' For millennia man sought that faith outside himself, in a certain view of the world provided by religion. This position has been greatly eroded for many by the advance of science, particularly by the Copernican and Darwinian revolutions. I believe that there is a great deal of strength to be discovered by looking inside ourselves, with the knowledge that each of us belongs to the same highly successful and diverse species. Our unusually developed degree of intelligence gives us a global perspective not available to other creatures. We should be able to appreciate that a balance of nature exists while at the same time judiciously exploiting what nature has to offer. With our global perspective comes a global responsibility, a responsibility both for the other members of the human race and for the many forms of life with which we share our world.

The lesson of Hiroshima must not be forgotten. We are technically capable of obliterating life on earth, but we are not, I firmly believe, instinctively programmed so that such a fate is inevitable. *Homo sapiens sapiens* is neither innately aggressive nor innately peaceable. The human brain is an extraordinarily adaptable and flexible piece of machinery. The dramatic transitions that people make when the industrialized world impinges on previously isolated and technologically simple societies testifies to this. Within a generation individuals can step out of a world of hunting-and-gathering and into the technically complex environment of modern society.

Unlike our forebears who became extinct, we are an animal capable of almost limitless choice. The problem facing us today is our inability to recognize the fact that we *are* able to choose our future. Many people are content to leave their future to the will of God, but I believe this is a dangerous philosophy if it avoids the issue of our responsibility. It is my con-

The devastated city of Hiroshima after the dropping of the atomic bomb in 1945.

viction that our future as a species is in our hands and ours only: I would remind those who rely on God's mercy and wisdom of the old adage 'God helps those who help themselves'. We must see the dangers and the problems and so chart a course that will ensure our continued survival.

For me the search for our ancestors has provided a source of hope. We share our heritage and we share our future. With an unparalleled ability to choose our destiny, I know that global catastrophe at our own hands is not inevitable.

The choice is ours.

Acknowledgments

I am particularly grateful to Roger Lewin, a friend and collaborator for many years, who has helped me throughout the preparation of this book. Without this help I don't believe the book could have been produced in time to coincide with the television series. Peter Kain is also owed special praise for his efforts in taking photographs in many parts of the world, often under quite difficult circumstances.

The history of this book is tied directly to the film series and I must express my appreciation to Graham Massey and Peter Spry-Leverton with whom the series was conceived and produced. Both of them, together with the others on the production team and film crew, have made the whole project a pleasure to be involved with and I am sincerely grateful. Similarly I owe a debt of gratitude to Linda Gamlin who edited the various phases of this manuscript and to Martin Bristow for the excellent design of the book.

During the course of filming the series and writing the book, many people gave particular help both in arranging things and in giving me specific information. I must mention Kamoya Kimeu, Glynn Isaac and Yves Coppens in particular along with Richard Lee, Alan Walker, Pat Shipman, David Pilbeam, my mother Mary Leakey, Barbara Isaac, Woo Ju-Kang, Henri and Marie Antoinette de Lumley, Roger Fouts, Erik Trinkhaus, Paul Bahn, Bob Brain, Elizabeth Vrba, Don Johanson, Philip Tobias, Kathy Schick, Nick Toth, Henry Bunn, Larry Keeley, Patricia Draper, Ralph Holloway, Alexander Marshack, Ralph Solecki, Margaret Conkey, Richard Klein, Stephen Jay Gould, Peter Williamson, Diane Gifford, Kay Behrensmeyer, Martin Pickford, Michael Day, Clifford Jolly, Owen Lovejoy, Alun Hughes, Ellen Kroll, Jack Harlan, Charles Nelson, John Kimengich, Gordon Hillman, Richard MacNeish, Kent Flannery, Michael Moseley, Judy Maguire, Joyce Marcus, Robert Feldman, Richard Ford, Andrew Hill, John Thackray, Bernard Wood, Ian Reid, Lynne Frostick, Karl Butzer, Aris Pouliannos and Ron Bowen.

A project such as the television series and this book requires very extensive contacts and consultations, some of which were made in my name but without my actual participation. It is inevitable that many people who have given help in some way or another will not have been mentioned specifically but they are all thanked sincerely. Similarly, there are a great many projects, issues and ideas that have not been mentioned in the film or the book and I would simply point out that the subject is too large for an exhaustive review. I have tried to present a popular account of an exciting, international, multidisciplinary endeavour and I hope that my many colleagues will bear with me over any inadequacies.

Finally, I would like to pay tribute to the Government of Kenya and the Museum Trustees who gave me the time to participate in the filming of the series.

Even with this time and with all the help that I have mentioned, I could not have achieved anything without the constant support, encouragement, tolerance and help of my family, particularly my wife to whom I am forever grateful. Without Meave, neither the film nor this book could ever have been attempted.

The Making of Mankind

Senior Producer	**Graham Massey**
Producer	**Peter Spry-Leverton**
Research	**Jane Callander** Neil Harris
Producer's Assistants	**Philippa Copp** Jane Ashford Joanna Hatley
Cameramen	**Alec Curtis** John Record
Sound Recordist	**Rodney Bond**
Film Editors	**Keith Raven** **Christopher Woolley** Paul Munn Chuma Ukpabi
Graphics Designer	**Peter Clayton**
Music	**Peter Howell**
Special Effects	**Steve Drewett**
Make Up	**Lisa Westcott**

Illustration Sources

10 Mary Evans Picture Library
11 Bob Campbell
12 Rodney Bond
14 John Reader/Colorific
17 Bob Campbell
23 Yves Coppens
24 Radio Times Hulton Picture Library
27 Ian Murphy
28 H. Reinhard/Bruce Coleman Inc.
30 Anthony Maynard
32–3 Nicholas Hall
36 Andrew Hill
38–9 Nicholas Hall
41 Bob Campbell
42 Rodney Bond
43 Rod Williams/Bruce Coleman Ltd
46 Eugene Fleury
47 David Pilbeam
48 Eugene Fleury
50 *Photograph:* David Pilbeam; *Artwork:* Ron Bowen
53 W. Garst/Tom Stack Associates
56 *Left:* The Illustrated London News; *Right:* British Museum (Natural History)
59 Nicholas Hall, after diagrams by Bob Brain
64 Bob Campbell/Bruce Coleman Ltd
66 Rodney Bond
67 The Cleveland Museum of Natural History
68 David Brill, © National Geographic Society
69 David Brill, © National Geographic Society
72–3 Ron Bowen
85 Timothy Ransom
89 Richard Wrangham/Anthrophoto
90–1 Ron Bowen
93 Edward Ross
96 Irven DeVore/Anthrophoto
98 Eugene Fleury
100 Richard Lee
101 Irven DeVore/Anthrophoto
102 Edward Ross
103 Edward Ross
104 Irven DeVore/Anthrophoto
106 Edward Ross
113 Eugene Fleury
119 *Top and bottom:* Glynn Isaac
121 Institute of Palaeontology, Peking
126 Rodney Bond
128 Rodney Bond
133 Joyce Tuhill
135 Michael Holford
136 Michael Holford
138 Keith Preston
140 Irven DeVore/Anthrophoto
147 Musée de l'Homme, Paris
148 *Top:* Joyce Tuhill; *Bottom:* The Illustrated London News
149 Eugene Fleury
151 Michael Holford
154–5 Ron Bowen
161 Jean Vertut
162 Jean Vertut
166 Michael Holford
168 Eugene Fleury
170 Jean Vertut
171 Jean Vertut
174 Begouën Collection, photo: Jean Vertut
176 Martin Bronkhorst, after The Abbé Breuil
177 Jean Vertut
178 Joyce Tuhill, after Alexander Marshack
179 Jean Vertut
182 Keith Preston
185 Bryan and Cherry Alexander
186 Michael Holford
190 Heather Angel
191 Eugene Fleury
192 Keith Preston
193 Keith Preston
194 Musée des Antiquités Nationales, Paris
196 Eugene Fleury
199 Robert Feldman, Field Museum, Chicago
201 Eugene Fleury
206 Eugene Fleury
208–9 Ron Bowen
213 Eugene Fleury
215 Michael Moseley/Anthrophoto
220 Michael Day
222 *Top:* The Illustrated London News
226 Andrew Hill
228 Michael Holford
232 Eugene Fleury
233 Michael Freeman
234 Michael Freeman
239 Keystone Press Agency Ltd
241 Irven DeVore/Anthrophoto
244 Keystone Press Agency Ltd
246 Keystone Press Agency Ltd

The photographs on the following pages were taken by Peter Kain (© Richard Leakey): **8, 16, 19, 34, 44, 54, 57, 61, 62, 77, 79, 80, 82, 83, 87, 111, 115, 116, 122** (top and bottom), **125, 132, 134** (left and right), **137, 143, 144, 146, 157, 158, 164, 181, 188, 203, 207, 211, 216, 218, 222** (bottom), **223, 231, 236.** The photographs on the jacket and in the opening pages of the book were also taken by Peter Kain.

Index

Numbers in *italic* indicate an illustration; numbers in **bold type** indicate the main entry on the subject.

Abell, Paul, 15, 81
Aborigines, 169, *201*
Abri du Souci, 192, *196*
Abri Pataud 191, *196*
Abu Hureya 206–11, *206*
Acheulean industry, 84, **117**, *134*, 134–35, 150–51
adaptation, **26**, 29, 45, 95, 241–42
Afar Triangle, 33
Africa, prehistoric sites of, 13–15, 33, 37, *49*
aggression:
  animal, 240–41
  human, 92, 219–25, 230–37, 240–45
Agricultural Revolution, 20–21, 184, **200–15**, 230
  climate change and, 200–03
  cultural evolution and, 204–05
  in New World, 212–15
  population growth and, 203–04
agriculture, 97, 99, 201–02
  and aggression, 229
  and possessiveness, 226–27, *228*, 229
  development of, 20, 196, 200–01, *201*, *203*, *208–09*, **205–11**, **212–15**
  productivity of, 203
Altamira, 163, **165–67**, *166*, *168*, 169, 172
Altxerri, *168*, *192*
Ambrona, 118–20, *196*
Amerindians, 198–200, 212–19
  hunting-and-gathering tribes of, *201*
  irrigation by, *199*, 212, 217, 235
  migration of, 159, *213*
  see also Mesoamerica; Peru
Arago cave, 145, *149*
Ardrey, Robert, 92, **221–22**, 225
Ariège Valley, 175, *188*
australopithecines, **66–67**, *67*, 69, 70, *72*
  and bipedalism, 71
  brain size of, 75
  diet of, 71–74
  extinction of, 94–95
  gracile, *see Australopithecus africanus*
  robust, *see Australopithecus boisei* and *Australopithecus robustus*
Australopithecus afarensis, 69–70
  brain size of, 70
Australopithecus africanus, **57–63**, 66, *72*, 221, *222*, 222–23, *223*
  diet of, 71–75
  'Mrs Ples', 64, *223*
  Taung child, *57*, 224
  see also australopithecines
Australopithecus boisei, *16*, 65–66, *66*, *72*
  diet of, 71–75
  'Dear Boy', 65, *66*
  'Nutcracker Man', 65, *66*
  see also Australopithecus robustus; australopithecines
Australopithecus robustus, **58**, *61*, 63, 65, *72*
  diet of, 71–75
  see also Australopithecus boisei; australopithecines
Awash River, *49*, 67, *68*
Aztecs, 234–35, *232*

baboon:
  competition with australopithecines, 95
  feeding habits of, *53*, *85*, 89
  'yawn threat' of, 240–41, *241*
Bahn, Paul, **184**, **189–92**, 194, 195–96
Bakate River, 37, *38–39*
  bâtons de commandement, *178*, 194–95
Beagle H.M.S., 25
Begouën, Count Henri, 172–73, 175
Begouën sons, 172–74
Bender, Barbara, 204–05
Bering Straits, 159, *213*
Behrensmeyer, Kay, 36
biochemical evidence for relationship between man and chimpanzee 10–11, 48–50
bipedalism, 42, **50–52**, *53*, 70–71
birth control, *104*, **105**, 203, 229
Black, Davidson, **120**, 123
bone, **62–63**, 79, 86
  cut marks on, *87*
  damage to, 64, 223–25
  destruction of, *36*
  fossilization of, *36*, 63–64
  tools, 151, 186, *186*
Border Cave, *149*, 158–59
Bordes, François, **124**, 137, 151, 152, 165
Boule, Marcellin, 147–48
Braidwood, Robert, 204–05
Brain, Bob, *54*, **58–60**, 62, 223–25
brain size, 52–53, 70, 111–12, 114, 118, 130, **131**, *133*, 144, *148*
brain structure, **130**, **131–33**, *133*
breccia, *59*, *62*, 62–63
Breuil, Abbé, **170**, *176*, 193
Broca's area, 132–33
Broken Hill, 145, *149*
Broom, Robert, 11, 57–58
Bunn, Henry, 86
burials, 152–53, *154–55*
Bushmen, *see* G/wi; !Kung; San

Campbell, Bob, 15
Cartailhac, Emile, **167**, 169
Catal Huyuk, 212
Catastrophe Theory, 22
cattle, 196–97
Cavalli-Sforza, Luigi, 212
cave art, 138, **160–64**, *162*, *164*, 165–72, *166*, *170*, *171*, *176*, *177*, *179*, *182*, 182–83, *192*, *193*
  composition in, 167
  end of, 183
  horse in, *162*, 167, 170–71, *171*, 178–79, *179*, 193–94, *193*, *194*
  reindeer in, 167, 169, **170**, 192, 193
  symbolism in, *138*, **170–72**, *171*, *176*, 182–83
  see also Ice Age art; portable art
cereals, **202**, *203*, 205–11, *207*
  see also einkorn; wheat
Cerro Sechin, *218*, **219**, 235, 237
Chancayillo, *236*, 237
Chan Chan, 198, *215*
Chew Bahir, *38–39*
Childe, Gordon, 202
chimpanzee:
  language abilities of, 127–30, *127*, *128*
  brain size of, *133*
  feeding habits of, 43, *89*, 89

relationship to man, 43, **48–50**
Chimu, **198**, *199*, *215*, 237
Chomsky, Noam, 129
Choukoutien, *113*, 120–23, *122*
finds from, 120, 121–23, *122*
cleavers, *see* tools
climate, *32–33*, 200–03, 206–07, 213
Cohen, Mark, 203–04
Combe Grenal, *149*, 151
'commanders' maces', *178*, 194–95
Conkey, Margaret, **172**, 175
continental drift, 31–32, *32–33*
Copernicus, Nicolaus, 10
Coppens, Yves, **64**, 67, 69
Creation, the, 22
Cro-Magnon Man, 147–48
see also Homo sapiens sapiens; man
Curtis, Garnis, 114
Cuvier, Baron Georges, **22**, 24

Dart, Raymond, **11**, *11*, 13, **56–68**, 63, 221–25
Darwin, Charles, **10**, *10*, 13, 24, *24*, **25–31**
Dawson, Charles, 55
'Dear Boy', 65, *66*
see also Australopithecus boisei
diet, 50–52, 71–74, 83, 86–88, 93–94, 214
see also food sharing; meat eating; plant food
Diluvial Theory, 22
division of labour, sexual, 94, **104–05**, 227

Dobzhansky, Theodosius, 21, 245
domestication of animals, 187–97, *196*
see also cattle; horse; reindeer
Draper, Patricia, 228–29
dryopithecines, *44*, 45, *46*
Dubois, Eugene, 112–14

earth, age of the, **20**, 22–23
East African Rift Valley, **48**, *49*, 64
einkorn, 209–10
see also wheat
Eldredge, Niles, 30
endocasts, *132*
evolution of living organisms, 20, **24–31**

'family group', 67, *69*
Fertile Crescent, 201, 202, 204, *206*, **206–12**, *207*
fire, use of, 121, 222
fishing, 200, *215*, **215–17**, *216*
Fitzroy, Captain Robert, 25
flake tools, *see* tools
Flannery, Kent, **204**, 212–13, 214–15
food sharing, 88–89, *90–91*, 92–94, 106–07, 226
'food-sharing' hypothesis, 93–94
Fort Ternan, 48, *49*
fossil record, **29–31**, 35–37, 43, 55, 58–64
'fossil void', 43
Fouts, Roger, **128**, *128*, 129–30, 139
Freud, Sigmund, 220–21, 238
Fuhlrott, Johann Carl, 146

Galapagos Islands, 25
'Garden of Eden' theory, 156, **158–59**
Gargas, *168*, 175–76
Gardner, Allen and Beatrice, 127–28
gathering, **92–94**, 100, 105, 107–08
'gathering' hypothesis, 92–93
genes, **26**, 29, 240
geology, 22–23, 31–35
'gesture' theory of language, 139
giant baboon, 118, *119*
Gifford, Diane, 35–36
Gigantopithecus, 46–48
Gould, Stephen Jay, **30–31**, 55, 71
gracile australopithecines, *see Australopithecus africanus*
gradualism, *30*, 30–31
G/wi, *93*, *98*, 100, *102*, *103*, 115
see also San

Hadar, *49*, 63–64, **67–70**, *69*
tools from, 78
handaxes, *see* tools
Harlan, Jack, 207
Harris, John, 15, 78, 81
heart urchins, *30*
Henslow, J S, 25
Hillman, Gordon, **202**, 210
Hiroshima, 238, 246, *246*
Hobbes, Thomas, 97, 101
Holloway, Ralph, **131–33**, 139
hominids, **11**, 15, 17–18, 43, 50–52, 55, 57–63, 66–67, 70–71, 78
ancestor of, 45, **48**, 70
language abilities of, 138–39
brain size of, 133
diet of, 43, 51–52, 71–75, 93
footprints of, 37, 40, *41*, *42*
hominoids, **44**, 45–48, 50–53
Homo erectus, 20, *90–91*, **110–25**, *111*, *113*, *115*, *116*, *119*, *122*, *134*, 142, 145, 156–57, *196*
and food sharing, *90–91*, 112
brain size of, 111–12, 114, **118**, 131
diet of, 75, *115*
tools of, 84, **117**, *119*, *134*, 134–35, 150–51
Homo habilis, 14, 17–18, *19*, 20, **65–66**, 70, *73*
brain size of, 66, 75, 112, **131**
diet of, 74–75
see also skull 1470
Homo sapiens, **18**, 42, 71, 148, 149
see also Homo sapiens sapiens; Homo sapiens neanderthalensis; man
Homo sapiens neanderthalensis, 142, **145**, 149
see also Neandertal Man
Homo sapiens sapiens, **18**, 20, 127, 205, *208–09*, 245
and speech, 18–20, 127
brain size of, 131, *133*, *148*
origins of, 145, **156–57**
tools of, *136*, 184–86, *186*
see also man
Homo sapiens soloensis, 145
horse:
domestication of, 193–95, *193*, *194*, *196*
images of, 167, 170–71, 193–94
systematic hunting of, 189
tooth wear of, 184, 193, 195
Howell, Clark, 64
Hughes, Alun, **58**, 60–61, *62*
hunter-gatherers, 86, 88–89, 92, *93*, 94, *96*, *98*, **97–109**, *100*, *101*, *102*, *103*, 125, 165, 184, 196, 200, 201–02, 206–07
birth control among, **105**, *105*, 203, 229
present-day distribution of, *201*
settling down of, 225–29
size of foraging bands of, 99
suckling by, *105*
women among, 225–29
hunting, **92**, 100, 105, *106*, 107–08, 115, 118, 186–92
hunting-and-gathering, *see* hunter-gatherers
'hunting' hypothesis, 91–92
'hunting magic', *170*, **169–71**, *176*, 177
Hutton, James, 22
hyaenas, 225, *226*

ibex, 188–89
Ice Age, *33*, 164–65, *191*, 206–07
end of, 183, 201–02, 213
Ice Age art, 160–83, *168*
end of, 169, 183
fish in, *192*
handprints in, 175, *177*
human form in, 179–82, *181*, *182*
'hunting magic' theory of, *170*, 169–71, *176*, 177
meaning of, 169–72, 176–79
see also cave art; portable art; symbolism
impala, *27*
Incas, 198
irrigation, 198, *198*, 212, 217, 235
Isaac, Glynn, **77–78**, **81–82**, 88, 93–94, 99, 134–35, 140

Jebel Qafzeh, *149*, 158
Jericho, *206*, 212, 230, *231*
Johanson, Don, **63–64**, 67–70, *69*
Jolly, Clifford, 50–52

Kalahari Desert, 98–99
Kanapoi, *49*, 67
Karacadag, 207
Keeley, Larry, 87–88
Keith, Sir Arthur, 56–57
Kimeu, Kamoya, *14*, **15**, 37–39, 110
Klasies cave, *49*, 186–87
Klein, Richard, 186–87
Koobi Fora, **15–17**, 37, *38–39*, *49*, *80*, 81
see also Turkana, Lake
Kroll, Ellen, **83**, 84–85
Kromdraai cave, *49*, 60, 61

finds from, 57–58, 63
!Kung, *97*, *98*, **98–109**, *100*, *101*, *105*, *140*
settling down of, 225–29
birth control among, 105, *105*, 203, 229
women among, 225–29
see also San; hunter-gatherers

La Chapelle-aux-Saints, 147–48, *147*, *149*
Laetoli, 40, *41*, *42*, *49*, 67, 69, 70, 78, 159
La Ferrassie, *149*, 153
La Marche, 179–80, 193–94, *196*
Lamarck, Jean-Baptiste de, 24–25
Laming-Emperaire, Annette, 170–71
La Mouthe, 167, *168*
land ownership, **226–27**, *228*, 229
language, 18–20, **127–41**, 242
and art, 136–38
and the brain, 131–33
in chimpanzees, 127–30
and tools, **134–36**, 139
'gesture' theory of, 139
Lapps, *185*, 187, 192
La Quina, 137, 195, *196*
Lascaux, *161*, **160–63**, *162*, 167, *168*, 170, *171*
Laugerie-Basse, *168*, 193, *196*
La Vache, 177, **187–89**, *188*, *196*
Leakey, Jonathan, 9, 13, 65–66
Leakey, Louis, 9, *11*, **11–14**, 17–18, *64*
and 'Dear Boy', 65
and 'Piltdown Man', 56
Leakey, Margaret, 15
Leakey, Mary, 9, **13–15**, 40, *41*, *64*, 65–66, *66*
Leakey, Meave, 15–17, *17*, 110
Leakey, Philip, 9, 13
Leakey, Richard, *8*, 9, 13, *14*, 14–18, *64*, 110, *146*
Lee, Richard, **94**, 98–99, 103–07, 109, 201, 226–29
Le Moustier, *149*, **150**, 152–53
leopard, 60, *61*
Le Placard, 195, *196*
Leroi-Gourhan, Arlette, 153
Leroi-Gourhan, André, **169**, 170–71
Lespugue caves, 192, *196*
Lespugue, Venus of, 180, *181*
Les Trois Frères, *168*, 172–75, *176*
Le Tuc d'Audoubert, *168*, 172–73, *174*
Levallois technique, 135, 150–51, *151*
limestone caves, 58, *59*, 165, 167
Lorblanchet, Michel, 179
Lothagam, *49*, 67
Lorenz, Konrad, 222
Lovejoy, Owen, **52**, 69–70, 71
'Lucy', 67, *67*
Lukenya Hill, 197
Lumley, Henri de, **124**, 145
Lyell, Sir Charles, 22–23

MacNeish, Richard, **202**, 213–14
Maguire, Judy, 225
Makapansgat cave, *49*, 58, 61, *220*, 221–25, *222*
Malta (Siberia), 180–82
Malthus, Thomas, 25
man:
and apes, 43
and speech, 18–20, **127–41**
and technology, 18, 20–21
as cultural animal, 241–43
as 'killer-ape', 21, 53, 221–22, 242
body hair of, 18
brain size of, 18, **131**, *133*, *148*
future of, 238–47
nose of, 18
skin colour of, 116–17
see also Homo sapiens sapiens; Neandertal Man
Marcus, Joyce, **233–34**, 235
Marshack, Alexander, **136–38**, 176–79, 182–83
Marshall, Lorna, 106–07
Martin, Henri, 195
Marx, Karl, 27
Mas-d'Azil cave, 192, *196*
Maya, 233–34
Mayr, Ernst, 29, 114
meat eating, 83, *85*, 87, *87*, 88, **93–94**, 115, 214
Mesoamerica, 201, 202, 204, **212–15**, 219, **230–35**, *232*, *234*, *236*
Mexico, *see* Mesoamerica
Middle East, *see* Fertile Crescent
migrations of early man, 112, 114, 156, 159, 189–92, 212, *213*
Moche, 198, *198*
modern man, *see Homo sapiens sapiens*; man
Monpazier, Venus from, *168*, 180
Monte Albán, 215, **230–35**, *232*, *233*, *234*
Montgaudier baton, *168*, 177–78, *178*
Moore, Andrew, 207–10
Moseley, Michael, 215–17
Mount Carmel, 157–58, *157*, *158*
Mousterian industry, 135, *135*, **150–52**, *151*
'Mrs Ples', 64, *223*
see also Australopithecus africanus
Mugharet es Skhul, *149*, 157–58, *158*
Mugharet et Tabun, *149*, 157–58, *157*

National Geographic Society, 14–15
natural selection, 26–29, *27*, *29*
Neandertal Man, 142, 145–59, *146*, *147*, *148*, **148–49**, *153*, 154–55
anatomy of, 148–50
brain size of, *148*, 148–49
cultural separation in, 151–52
extinction of, 156–59
map, *149*
ritual burials of, 152–53, *154–55*
tools of, *see* Mousterian industry
'Neandertal phase' theory, 156–59
Nelson, Charles, 196–97
Nelson Bay cave, *49*, 187
Ngeneo, Bernard, 17
Niaux, *168*, 170
North American Indians, *see* Amerindians
nuclear warfare, 238–40, *246*
'Nutcracker Man', 65, *66*
see also Australopithecus boisei
Nzube, Peter, 15

Oaxaca Valley, **214**, **230–32**, *232*
Olduvai Gorge, *12*, **13**, 15, 33, *49*, 64, 65, 81, *87*
Olorgesailie, *49*, 118, *119*
Omo Valley, **14**, *23*, **33**, *49*, 64, 78, 159
Origin of Species, The, 25–30
'osteodontokeratic' culture, *222*, 223–24

Pangaea, 31–32, *32–33*
pastoralism, 97, 196–97
Pech de l'Azé, 137
Pech-Merle, *168*, *177*, 178–79, *179*
Pei Wen-chung, 120
Peking Man, 120–23, *121*, *123*
Peru, 212
coastal plain of, **198–200**, *199*, 215–19, 235–37
Petralona, 142, *143*, *149*
Petralona skull, **142**, *144*, 145
phyletic gradualism, *30*, 30–31
Pickford, Martin, 46
Piette, Edouard, **165**, 167, 193
Pilbeam, David, **32–33**, 43, 45, 46–48, 50, 52–53, 55, 241
'Piltdown Man', 13, 52–53, **55–56**, *56*
Pithecanthropus alalus, 113–14
Pithecanthropus erectus, 113–14
see also Homo erectus
plant foods, 86–87, 88, 92, 93–94, 206–10, 214
population, 200, 212, 243
portable art, 136, 137, **168–69**, 172, 173, *178*, *181*, *194*
see also cave art; Ice Age art
poverty, *238*, 243–45
primates, 42
feeding habits of, 43, 52
Proconsul africanus, *44*, 45
public works, 217, 230, 233, 235–36
punctuated equilibrium, 31

ramapithecines, 45–48, *46*, *47*, *50*
Ramapithecus, **46–50**, *50*, 52, 55, 70
diet of, 74–75
Reck, Hans, 13
reindeer, *185*, 187, **189–92**, *190*
domestication of, *185*, 192–93
Ice Age migrations of, **189–90**, *191*, 192, *196*
images of, 167, 169, 170, 192, 193, *196*
religion, 10–11, 245
Rhodesian Man, 145
Rift Valley, **48**, *49*, 64
Robinson, John, 74
robust australopithecines, *see Australopithecus robustus* and

Australopithecus boisei
rock paintings, *see* cave art

Sahlins, Marshall, **98**, 219
St Acheul, *113*, 117
St-Michel d'Arudy, 193, *194*, *196*
Salasun, 197
Salinas de Chao, 215–17, *216*
San, 86, *98*
see also G/wi; !Kung; hunter-gatherers
Sarich, Vincent, 48–50
Sautuola, Marcellino de, 165–67
Schick, Kathy, 85
Sechin Alto, 235–37
sedimentary rocks, *23*, **30**, 33–35, *38–39*
selection, *see* natural selection
Shah, Ibrahim, 45
Shanidar, *149*, 153, *154–55*
simple tools, *see* tools
site 50, *76*, **81–88**, *82*, *90–91*, 99
Sivapithecus, 46–48, *47*
Siwalik Hills, **32–33**, *34*, 45–48
skin colour, 116–17
skull 1470, *16*, **17–18**, *18*, 133
see also Homo habilis
skulls, damage to, 221, *223*, **223–24**
Slocum, Sally, 92
Social Darwinism, 26–27
Solecki, Ralph, 153
Solo River, 113–14, *149*
Solutré, 189, *196*
'sorcerer', 174–75, *176*
South Africa, 15, 56–63
speciation, **30–31**, 156–57
Spencer, Herbert, 26–27
Steinheim skull, 144–45, *149*
Sterkfontein, *49*, 57–58, 60–61
finds from, 57–58, 63, *223*
stone tools, *see* tools
'struggle for existence', 27
Sturdy, Derek, 189–90
'survival of the fittest', 26–27
Swanscombe skull, 144, *149*
Swartkrans cave, *49*, *55*, 58–61, *59*
finds from, 58, *61*, 63, *132*
symbolism, 137–38, 170–72, 176–79, 180–83

Taieb, Maurice, **63–64**, 67, 69
Tanner, Nancy, 92–93
Taung, *49*, 56–58, 63
child, *57*, 224
see also Australopithecus africanus
Tehuacan Valley, 202, **213–15**, *232*
Tello, Julio, **219**, 235
Teotihuacán, 230–32, *232*, **234–35**
Terra Amata, *113*, 123–34, *125*
Teshik Tash, *149*, 153
Thant, U, 243
Tibiran, *168*, 175–6
Tobias, Phillip, 60–61, *64*
tools, 76–78, 84, 93, 185–86
and language, 134–36
bone, 151, 186, *186*
cleavers, 84
flakes, 78, **83**, 84, 86, 124
handaxes, 13, 84, *119*, *134*
manufacture of, *83*, 135–36, 151
of modern man, *136*, 184–86, *186*
sickles, *207*
simple tools, *85*, *134*
stone, 61, **78–85**, *85*, 86–88, 112, 125, *134*, 134–37, *135*, *136*, *151*, 186
see also Acheulean industry; Levallois technique; Mousterian industry
tooth wear, **74–75**, 115, *115*
of horses, 184, 187, 193, **195**
Torralba, 118, *196*
Toth, Nick, *83*, 83–84
trade, 212, 230
trance-dancing, 99–100, *100*
travertine, *59*
tree shrew, *43*
Trinil, *113*, 113–14
Turkana, Lake, **14–17**, *49*, 66–67, 78
formation of sediments, 35–39, *38–39*
see also Koobi Fora

Uncle Philipos, 142, *143*
Ussher, Archbishop James, 22

variation, **26**, 29
Venuses, 180, *181*
Venus of Lespugue, 180, *181*
Vogelherd figures, **136–37**, *137*, 178–79
Vrba, Elizabeth, **60–61**, 63

Walker, Alan, **17**, *17*, 31, 74–75, 76, 114–15, *115*
Wallace, Alfred Russel, **26**, 112
war, *see* aggression; nuclear warfare
Washburn, Sherwood, 92
Washoe, **127–29**, 130
waste of resources, 243–45, *244*
Weidenreich, Franz, 123
Wernicke's area, 132
see also language
wheat, *203*, 206, *208–09*, 209–11
White, Tim, 69, 70
Whorf, Benjamin Lee, 127
Wilson, Allan, 48–50
Woo Ju-Kang, *122*, 123
Wood, Bernard, 15

Zapotec, *232*, 235
Zihlman, Adrienne, 92–93

Glossary

basalt: a dark-coloured rock, formed by the solidification of lava from volcanoes.

breccia: a rock formed in caves and composed of fragments of other rock, mixed with soil, silt and sometimes bones.

endocast: a cast of the inside surface of a skull.

fault: a crack in the earth's crust along which there has been movement.

hominid: all members of the family Hominidae.

Hominidae: the family which includes all *Homo* and *Australopithecus* species, but excludes the apes.

hominoid: a collective term for the apes and hominids.

natural selection: a theory which aims to explain the mechanism of evolution. It proposes that the natural variation found in a population leads to greater or lesser chances of survival for different individuals. This in turn affects the extent to which their genes are passed on to the next generation.

Neolithic: the 'New Stone Age'; the era which saw the beginnings of agriculture.

palaeoanthropology: the study of human ancestors.

Palaeolithic: the 'Old Stone Age'; the pre-agricultural era.

palaeontology: the study of animal and plant life in times past.

primates: the order of mammals which includes humans, apes and monkeys, as well as lemurs, pottos, bushbabies, lorises and tree shrews.

radiocarbon tests: the use of changes which occur naturally in carbon to determine the age of a fossil or other object. As organic material contains a great deal of carbon it is a useful test for biological remains. However, the changes are almost completed after 120,000 years, so a radiocarbon test cannot be applied to material older than this.

sedimentary rocks: rocks composed of sediment; the sediment itself is formed by the weathering of other rocks.

List of Maps

Sites of finds of ramapithecines and dryopithecines 46
Africa: the major prehistoric sites 49
The distribution of the !Kung and other San peoples 98
Major sites of finds of *Homo erectus* 113
Major sites of finds of Neandertals and other early forms of *Homo sapiens* 149
Major sites of finds of Ice Age art in France and northern Spain 168
Possible migration routes of reindeer herds in Ice Age Europe 191
Sites in France and Spain where evidence suggests the control of animals by early man 196
Present-day distribution of hunter-gatherer peoples 201
The Fertile Crescent 206
The land bridge between Asia and America during the Ice Age 213
Mesoamerica, 2,000 years ago 232